Smoke Wars

Smoke Wars

Anaconda Copper, Montana Air Pollution, and the Courts, 1890-1924

by Donald MacMillan

Smoke Wars: Anaconda Copper, Montana Air Pollution, and the Courts, 1890–1924

Book design by DD Dowden, Helena

Cover design by Kathryn Fehlig, Helena

Typeset in Adobe Garamond and Calligraph

Printed by Creasey Printing, Springfield, Illinois

P.O. 201201, Helena, MT, 59620-1201

00 01 02 03 04 05 9 8 7 6 5 4 3 2 1

Library of Congress Cataloging-in-Publication Data
MacMillan, Donald, 1932–
Smoke wars: Anaconda Copper, Montana air pollution, and the courts, 1820–1924/by Donald MacMillan ; introduction by
William L. Lang
p. cm.
Includes bibliographical references and index.
ISBN 0-917298-62-4 (cloth: alk. paper)
ISBN 0-917298-65-9 (pbk.: alk. paper)
1. Copper industry and trade—Environmental aspects—Montana.
2. Smelting—Environmental aspects—Montana. 3. Anaconda Copper Co. 4. Air—Pollution—Montana. 5. Environmental protection—Political aspects—Montana. I. Title
TD888.C64 M33 2000
363.738′4′09786—dc21 00-023061

To my wife "Philli"
and my children Lois Cowell, Shannon Taylor,
and Father Leonard MacMillan

This book was made possible in part
with funding from the Bair Trust.

Contents

– Maps –

– Illustrations –

– Introduction –

Bad Air, Environmental Politics, and History

by William L. Lang

In 1972 when Montanans talked over coffee cups and shot glasses about public affairs in their state, the subjects of coal mining, air quality, and water use provoked heated debates. Ever since publication of the North Central Power Study the previous year, the prospect of massive strip mining operations on Montana's eastern plains and the construction of coal-burning power plants at the mines had politicians, ranchers and businessmen, and environmentalists offering opinions on the coal issue. Air quality ranked high on the list of environmental concerns. The North Central Power Study had proposed that half of the planned forty-two coal-fired electricity generating plants on the northern plains be located in eastern Montana. Burning coal, even the low-sulfur coal of the Fort Union Formation, meant deterioration in air quality. Some pointedly asked if Montana really wanted to trade its vaunted Big Sky clean air for the economic rewards promised by mining the state's estimated 108 billion tons of coal and utilizing the consequent power generated.

A series of open town meetings offered the public opportunity to address the issue. In Helena, the clean air hearing drew an overflow crowd. Industry representatives promised that coal-fired power plants would not cause serious pollution, that the air would remain clean and not pose a health risk. Many remained skeptical, but industry representatives presented scientific testimony to support their claims and described the mitigation technologies they could afford and stood ready to introduce. But they also cautioned that overly restrictive legislation might cripple their ability to operate the power plants, and might even drive them out of business. It was a simple message that Montanans had heard before. Do not worry about pollution. Science will fix anything that might go wrong. Do not require too much protection, for that would slay the goose and Montana's economy would be the loser.

During the course of the meeting, a University of Montana scientist read a published statement that he warranted represented industry's viewpoint. The piece laid out a litany of objections to the statutory regulation of industrial emissions, arguing that technology could not prevent the emissions, that no adverse health effects had been documented, and that in any case the industries could not endure the extra expense. The audience nodded knowingly. The article appeared to contain more of the explanations and opinions that copper smelters, coal companies, and other producers of polluting emissions had already offered in defense of the new plants. But they stood in gaped shock when the scientist revealed that the article was not the work of Peabody Coal Company, ASARCO, or the Anaconda Company; metallurgical industry spokesmen in England had written it in the 1880s. The objections of industry to air quality standards were old, very old.

While the hearings grabbed headlines in the state, University of Montana historian Donald MacMillan worked on finishing his Ph.D. dissertation on the history of smoke abatement and litigation against mining and smelting industries in Montana, the so-called Anaconda smoke cases. MacMillan's dissertation, "A History of the Struggle to Abate Air Pollution from Copper Smelters of the Far West, 1885–1933," has been edited posthumously and is published here under the title *Smoke Wars*.

It is tautological to say that historians write from the perspective of their times, but the rule wages even more importantly in the case of Don MacMillan's investigation of the smoke litigations. Not only did his research present industry arguments that paralleled the protestations offered by the English smelting industry during the 1880s and the more modern arguments of the power industry in the 1970s, but it also followed a historiographical tradition in Montana that can best be described as profoundly skeptical of the state's corporate behemoths, especially the Anaconda Copper Mining Company.

MacMillan's dissertation advisor was K. Ross Toole, A. B. Hammond Professor of History at the University of Montana, former Director of the Montana Historical Society, and author of the pointedly argued *Montana: An Uncommon Land.* Toole looked at Montana's past from an anticorporatist and anticolonialist viewpoint. Heavily influenced by the writings of Walter Prescott Webb, Bernard DeVoto, and Joseph Kinsey Howard, Toole cast Montana history as a morality play that pitted the populist citizenry and its champions against the forces of external corporate power and its political operatives. His heroes stood like Davids, hurling sometimes effective political missiles against the Goliaths of the Anaconda Mining Company, Montana Power Company, and the Northern Pacific Railroad, but failing in the end to drop the giants to their knees.

In *Twentieth Century Montana: A State of Extremes*, published one year before MacMillan completed his dissertation, Toole used a series of vignettes to argue that the state had barely survived orchestrated corporate efforts to control the lives, pocketbooks, and minds of its citizens. Only the character and irrepressible optimism of Montana's people, Toole maintained, had kept the body politic from total capitulation to corporate forces. The connection between Montanans and their land, he claimed, was essentialist. "If the fight against environmental degradation can be won anywhere," he wrote, "it will be won here—precisely because nowhere in America is that visceral relationship with the land more powerfully felt by those who live here."[1]

Toole strode the academic halls in Missoula as a clarion for a populist interpretation of the state's past. His students adored his passion, his storytelling gifts, and his political activism, especially his support of environmental causes. The steady resistance to proposed environmental legislation in the state legislature came from natural resource industries whose capital and direction originated outside the state. As in the days of the Copper Kings, as Toole would have it, Montana appeared to be suffering from colonial depredations. Toole's message compelled and stimulated his graduate students, fueling some of the best monographic studies of the state's history. More important for MacMillan and the few others who pursued environmental topics, Toole's suspicions about corporate players in Montana's historical drama heavily informed their attitudes about corporate actions during the 1960s and 1970s. Some of the politicians who took up the cause of preserving the state's threatened environment had listened to Toole's lectures in History 367, "Montana and the West," and absorbed the anticolonialist message.

In the 1970s, Montana environmentalists often combined their skepticism about corporate political behavior with their concerns about air quality, water resources, and land use. Scraping off the relatively thin overburden on Montana's plains to gouge out low-sulfur coal worried everyone who advocated environmentally sound land use or doubted the success of land reclamation efforts. In 1968, Montana Power Company's subsidiary, Western Energy Company, began a new strip mining operation at Colstrip, Montana, where the Northern Pacific Railroad Company had initiated strip mining for coal in the 1920s.

Peabody Coal Company followed this lead and soon began shipping carloads of low-sulfur coal to Midwest and West Coast utilities. By 1969, Western Energy Company had announced plans to construct two 350-megawatt and two 700-megawatt power plants at Colstrip. The age of coal had come to eastern Montana. The Yellowstone River Basin's water resources became targets for massive development to provide power plants with thousands of acre-feet of impounded and readily available water. A proposed dam at Allenspur Gap, near Livingston on the upper Yellowstone, became an iconographic warning for committed environmentalists who rallied fly fishing organizations and other recreationists to resist industry proposals for coal and power developments in eastern Montana.[2]

Federal legislation that reflected a national concern for environmental quality created the Environmental Protection Agency in 1970 and a series of laws designed to protect air and water quality. Among the new laws was an act that set new standards for coal-fired electric plants and forced large Midwest utilities to purchase low-sulfur coal from Montana and Wyoming. Out on Montana's eastern plains, the coal rush shocked traditional ranching communities, and the prospect of huge electrical generating plants in their own backyards compelled local ranchers to form a coalition with liberal urbanites from western Montana. The Northern Plains Resource Council pledged to fight massive coal developments and preserve the land for stock raising and environmentally friendly industries. In their monthly publication, *The Plains Truth*, the Council informed environmentally concerned Montanans about the latest energy companies' plans to acquire and develop coal lands.

Resistance by eastern Montana ranchers dovetailed with efforts in western Montana to curb air and water pollution near paper pulp plants. Local grass-roots organizations, such as G.A.S.P (Gals Against Smog and Pollution) in northwestern Montana, fueled activist campaigns to change attitudes in the legislature. In Helena, environmentalists created the Environmental Information Center, which alerted Montanans to national environmental legislation that affected the state and aided in state efforts to regulate natural resource industries. Local chapters of The Nature Conservancy, the Sierra Club, and the Audubon Society added to their members and increased a groundswell for environmentally informed political action. Voters responded by sending a phalanx of young, liberal, committed legislators to Helena for three successive biennial sessions—1971, 1973, and 1975.[3]

Legislators passed a series of laws that created aggressive and responsive oversight of natural resource industries and other developments that potentially threatened the state's environment. For example, in 1971 the legislature created the Environmental Quality Council and, in 1973, passed a Utility Siting Act to control installations such as the power plants at Colstrip. By the end of the 1975 legislative session, Montana had laws on the books that required coal mining reclamation and re-registration of all water claims, placed a moratorium on development of Yellowstone River waters, and established the highest coal severance tax in the nation. Although there had been earlier discussions in the legislature about environmental issues ranging from fish and game laws to Indian water rights, it had taken the prospect of massive coal mining and power generation to drive the state to action.[4]

Lobbyists for industry and environmental groups argued their proposals before legislative committees that oversaw commerce, human health, finance, and other state concerns. These contests in the legislature informed political discourse in the media, in university classrooms, and elsewhere. Industry's unwillingness to accept the costs and responsibilities for additional environmental protection often sparked the most intense debates, in part because it appeared to be one more installment in a familiar story about Montana's natural resource wealth and the political conflicts it had triggered. The tale was the same—outside capital interests taking advantage of Montana with little to no regard for the future—only the dates, locations, and specific individuals were different. Those who believed the rumors of Anaconda's support for the 1973 coup in Chile that ousted Salvador Allende found an appropriate prologue in the story of the Company's famous mine shutdown in 1903, which had forced legislators and a governor to do the Company's bidding. There were other suggested parallels. The industrial conglomerate MacMillan studied—known at various times as the Amalgamated Copper Company and the Anaconda Copper Company—had been the result of Wall Street financial combinations directed by Standard Oil Company interests that had gobbled up the Anaconda Copper Mining Company. The impending corporate realignment during the 1970s would bring another oil titan—ARCO—to Montana as the new owner of the historic Anaconda properties in 1977.

Although profoundly influenced by the anticorporate and anti-Anaconda stands taken by many environmentalists during the 1970s, MacMillan did not write a polemic against the Company. He delivered strong opinions about the Company's intentions and behavior, but he drew those conclusions from a sizable body of evidence. MacMillan begins his story in the 1880s and 1890s with the first open challenge to Butte copper companies' practice of heap roasting ores—an open-air smelting practice that created dense clouds of low-lying, noxious smoke. Citizens had long complained of the smelting fires in Butte, but not until the civic improvement campaigns at the end of the nineteenth century stimulated citizens to reformist action did they bring suit to halt heap roasting. The companies resisted and then yielded, in a way, by transferring smelting to the great reduction works in Anaconda, twenty-six miles west of Butte.

The relocation of the nuisance, however, only created a new objection, this time from the farmers and ranchers in the Deer Lodge valley. The huge Anaconda smelter that Marcus Daly had located along Warm Springs Creek in 1884, smack against the mountains at the upper end of the Deer Lodge valley, boasted a perfect combination of plentiful water and ample flat land to build a series of mill buildings. The smelter smoke issued from a huge stack that measured three hundred feet high by 1903 and sent toxic fumes thirty miles north down the Deer Lodge valley, long a center for stock raising and farming operations. The smelter's toxic plume included high concentrations of sulphur dioxide (SO_2) and arsenic. The farmers and ranchers complained that SO_2 destroyed vegetation and arsenic sickened their stock. The U.S. Forest Service added complaints against the smoke for stunting the growth of trees in the national forest.

MacMillan details the litigation caused by suits brought against the Anaconda Copper Mining Company by the Deer Lodge valley agriculturalists, first in state court and then in federal court. The so-called "Anaconda smoke cases," MacMillan argues, were part of a larger campaign to control environmental damage from copper smelters across the West. The conclusions he draws are familiar to readers of late twentieth-century pollution cases: business interests get off with few penalties. In the Anaconda instance, as MacMillan explains, the Company promised mitigations to their smelter emissions in an out-of-court settlement, but they delivered on few of them, while they bought out Deer Lodge valley farmers and swapped forested lands with the federal government to compensate for losses to the public domain.

MacMillan's study of the Anaconda smoke cases stands as one of the earliest environmental histories in Montana historiography, but it is also a considered evaluation of hotly contested political and economic disputes. *Smoke Wars* is about more than politics. MacMillan approached his topic as an interdisciplinary work, by incorporating significant information from scientific findings in addition to economic, legal, and political studies. His focus on the unhealthy consequences of airborne chemical emissions from copper reduction processes, for example, forced him to investigate the scientific data presented by government and industry experts in court, data that often pitted scientist against scientist. In those disputations, scientists testified about the effects of SO_2 on grasses and the effects of arsenic on animals in the Deer Lodge valley. Other testimony outlined new technologies for eliminating significant portions of SO_2, arsenic, and other contaminants from smelter smoke. Legal discovery that included sophisticated scientific measurements and technologies, such as electrolytic precipitation, had not been part of case law for very long in 1910. Part of MacMillan's task, then, was to explain the effect of new science and technology on a case that pitted Montana's wealthiest corporate actors against aggrieved farmers and ranchers in the Deer Lodge valley, and later against the federal government.[5]

Although science and technology are important in MacMillan's study, the heart of his argument rests on environmental politics. The struggle MacMillan documents was between a set of citizen representatives, which ranged from Butte's "citizen smoke committee" to attorneys general under Theodore Roosevelt and William Howard Taft, and attorneys and spokesmen for the Amalgamated Copper Company, notably Cornelius "Con" Kelley. The arenas included public meetings, press stories and public relations campaigns, courtrooms, and several meeting rooms, where representatives of each side hammered out compromise agreements. MacMillan spares few critical modifiers in his characterizations of the Company's self-protective actions. He charges the Company with being heedless to the interests of other legitimate political groups and with pursuing its own economic self-interest at the expense of the environment. The scenario is comparable to Toole's descriptions of unsavory bargains between politicos in his

Twentieth-Century Montana and manipulations by coal companies in *Rape of the Great Plains*.

There is appeal in this argumentative strategy, but there is also weakness. When MacMillan wrote this study, environmental history had barely emerged as a new field. Universities offered the first environmental history courses during the early 1970s, and only a handful of monographs and biographies that could be called environmental histories had been published by 1973. Nonetheless, a significant body of material about air pollution could have given MacMillan additional perspectives on the Anaconda smoke cases. By 1915, more than a dozen U.S. cities had passed ordinances against smoke nuisances. Most of the laws categorized smoke in gradations of visible aspects, with some outlining more than ten degrees of color, from black and dark to gray and whitish atmospheric effects. In Philadelphia, the law specified "smoke intercepting more than 60 per cent of light, and fumes of sulphurous or noxious odors" as violations and subject to fines. These laws drew upon a history of antismoke legislation that rested on English and German efforts to curb coal smoke in the thirteenth and fourteenth centuries and on the first published criticism of industrial smoke emissions, John Evelyn's *Fumifugium*, published in 1661. Modern scientific investigations of the dangers of industrial smoke began in 1819 with a published study of London's smoky atmosphere.[6]

The type of smoke nuisance MacMillan documented in Butte had analogs in other industrial settings in the U.S. The dense smoke episodes created by heap roasting in Butte during the 1880s and 1890s created social and political conditions that other, more industrialized cities experienced at the same time. One case in point that might well have informed MacMillan's analysis of conditions in Butte and the Deer Lodge valley is the famous Pittsburgh Survey of 1909–1914, which evaluated living conditions in arguably the most industrialized city in the nation.

Several studies included in the survey described the effects of industrial smoke and discussed efforts by civic reformers and city officials to eliminate noxious emissions. The dense clouds of dark and foul-tasting smoke became increasingly objectionable to city residents during the last decades of the nineteenth century. Laws had been passed as early as 1868, but the economy of using fuels with high sulfur content during the 1880s and 1890s had made Pittsburgh very much like Butte. By the mid-1890s, citizen committees began protesting against offending industries. Their efforts forced officials to address the problem by establishing a smoke inspector's bureau in city government in 1907. But smoke abatement efforts faced severe limitations in Pittsburgh and other cities. Smoke emissions were considered nuisances, which limited enforcement and penalties to measurements of inconvenience rather than as health risks. Attempts to link smoke emissions to the deterioration of human health failed, while industries and municipalities agreed to implement overly optimistic and flawed mitigation programs.[7]

During the Progressive Era of political and social reform, scientists addressed environmental issues in forestry, urban water systems, and other areas as never before and often with government assistance. Officials generally included smoke abatement on the list of problems for mitigation and focused on new technologies that could eliminate particulates and chemical substances in industrial flues. By 1912, American scientists had adopted English techniques in monitoring atmospheric gases that used liquid precipitation techniques to measure chemical components. In 1883, J. A. Clark and Oliver Lodge experimented with ionization to precipitate dust particles in industrial flues. Between 1906 and 1912, Frederick Cottrell and Walter Schmidt developed two technologies that successfully applied high-voltage transformers to electrostatically precipitate SO_2 out of gaseous streams in flues. Electrostatic precipitation and other technologies, which MacMillan describes in his discussion of mitigation efforts at smelters in California and Colorado, were part of the new reliance on science and technology as solutions to environmental, political, and social problems in the new century.

In one of the earliest modern environmental histories, Samuel P. Hays comments on the new prominence of scientific and technological expertise during the Progressive Era. Hays argues that reformers and politicians relied uncritically on the magic of expertise, technology, and scientific efficiency to solve problems. In MacMillan's reading of the Anaconda cases, the promise of technology fails to prevail, but it is unclear why. There are a multitude of studies on the effects of toxic gases in areas near smelters in the West. The technologies offered as nostrums appeared to be promising at the least and possibly solutions. Yet, the scientific findings and new technologies became seeming discards in the litigious proceedings. MacMillan answers the question partly by emphasizing the Company's influence in the Anaconda Smoke Commission, which had responsibility for evaluating the claims against the Company and crafting some resolution to the dispute.[8]

He might well have addressed the question anew in the revision he planned to write. He died in 1996 before he could complete the work, but surely the development of the field of environmental history and new information and analysis about his subject by scientists and historians would have provided a richer interpretation of the events surrounding the Anaconda smoke cases. Foremost in new findings about air pollution since the early 1970s is the corpus of scientific studies. The Environmental Protection Agency, established in 1970, identified six criteria pollutants: sulfur oxides, particulates, carbon monoxide, photochemical oxidants, hydrocarbons, and nitrogen oxides. New scientific investigations used very sophisticated measurements to target compounds that incorporated these pollutants and might be considered carcinogens. Measurements that had been impossible when the Anaconda smoke cases were adjudicated became everyday scientific data through the use of the electron capture device, an instrument that can detect and measure compounds in parts per trillion in the atmosphere.[9]

Once the EPA established air quality standards that applied to industrial emissions factories and plants, coal-fired power plants and nonferrous metals smelters became among the first objects of monitoring surveys. In 1977, when amendments to the 1970 Clean Air Act became law, the EPA added visibility to the basic standards and allowed states to pass tougher standards in their clean air legislation, an option that Montana pursued. Throughout the 1970s, the EPA tried to establish specific health requirements for each of the six criteria pollutants to compare with toxicological and epidemiological studies. The controversies over the specific effects of SO_2 on vegetation that MacMillan described are repeated in kind in current disputes between industries and federal regulators. The complaints by polluting industries are similar, but the scientific bases of the disputes are far more technical and more testable. Copper companies and iron and steel makers have contested EPA standards in recent years, using some of the same categorical arguments Anaconda used in 1912.[10]

While the science of environmental studies flowered as practically no other field did during the two decades after the passage of the National Environmental Policy Act in late 1969, environmental history also developed at a rapid pace. The first studies focused on forest and land use policies, on the history of wilderness ideas and ethics, and on water rights and projects. By the mid-1980s, several scholarly journals addressed the subject and an academic association—the American Society for Environmental History—dedicated its efforts to promoting more sophisticated study of relationships between humans and the environment.[11] Urban environmental subjects, such as garbage disposal and toxic pollution, first attracted historians' attention during the early

1980s. Articles and monographs by Martin Melosi, Joel Tarr, Craig Colten, and other scholars have expanded our knowledge of the history of pollution abatement in America and sharpened our analyses of the contentious and often ideological characteristics of urban environmental disputes.[12]

The most important and extensive studies on the history of industrial air pollution focus on post-World War II conditions. Of particular relevance to MacMillan's study of the Anaconda cases are studies by Joel Tarr and Carl Zimring on Pittsburgh and St. Louis. During the first half of the twentieth century, coal burning caused most urban smoke problems, whether the source was industrial or residential use. In St. Louis, for example, the high-sulfur coals that lay east of the city became an inexpensive but polluting fuel of choice for St. Louis businesses and homeowners. By the early 1940s, the imperfect and unenforced smoke control legislation passed by the city and the state of Missouri between 1901 and 1924 had failed to solve the problem, prompting the creation of an aggressive citizens' smoke committee. Sometimes, as in the case of the famous Donora, Pennsylvania, air pollution disaster in 1948, where twenty-two people died of smoke inhalation, major, tragic occurrences triggered action. By the end of World War II in St. Louis, Chicago, Pittsburgh, and other cities, direct citizen action by middle-class voters who worried about public health had created a new force in the environmental movement, a force that Samuel P. Hays has argued is tied directly to the postwar burgeoning economy. As in the Anaconda cases, good economic times and increased industrial output brought complaints from groups who paid an unacceptable environmental price for general prosperity.[13]

Women led many of these grass-roots campaigns, as they had during the Progressive Era. In 1962 with publication of Rachel Carson's *Silent Spring*, however, the tenor of the grass-roots movement changed. Perhaps the most important environmental book in the twentieth century, *Silent*

Spring alerted citizens to a threat to human health that could not be intuitively perceived, the invisible dangers of man-made chemicals, including compounds such as DDT, high concentrations of SO_2, and airborne heavy metals. As historians Robert Gottlieb and Carolyn Merchant have argued, women joined effective grass-roots organizing with serious threats to human health to achieve legislation that put controls on pollution and castigated polluters. Although many groups brought suit against polluters, public relation campaigns and strident political pressure often brought faster results in state legislatures and Congress. Lois Gibbs's crusade against the polluters of Love Canal, New York, for instance, led directly to passage in 1980 of the Comprehensive Emergency Response, Compensation, and Liability Act, the so-called Superfund law. The former Anaconda Copper Mining Company properties in Butte—many that figured in MacMillan's study—came under Superfund designation in 1989.[14]

At the end of the twentieth century, environmentalists and environmental historians know a good deal more about industrial pollution than they did thirty years ago. Controversies continue to swirl around mining operations and the plans of mining companies. During the 1990s, for example, disputes erupted over planned mines on the Big Blackfoot River and on land just east of Yellowstone National Park. Anaconda and Butte operations have also come under scrutiny. As a result of the Superfund legislation in 1980, the U.S. Department of Justice listed the corridor between Butte and Anaconda as one of the most toxic of industrial places in the nation in the early 1980s. Because ARCO had purchased the assets and properties of the Anaconda Copper Mining Company in 1977, the oil giant also gained the Superfund liabilities of the Anaconda properties. Litigation initiated by the federal government in 1989 charges ARCO with cleanup and mitigation responsibilities under the Superfund law. The company is contesting the government's suit and has filed countersuits claiming that the federal government's executive departments are partially responsible for existing toxic pollutants. The cases have brought the old issue that MacMillan documented back into public debate. Adjudication of the issues, once again, partially rests on scientific reports about the causes and consequences of dispersing industrial pollutants into land, water, and air.[15]

The tactics used by Anaconda and other corporations to avoid responsibility for pollution damages continue to be used. The "smoke rights" that the Company gained in 1910 from Deer Lodge valley farmers and ranchers whose lands lay within the smoke zone of the Anaconda stack, for example, are somewhat analogous to the recent idea that pollution rights can be purchased and traded among polities to encourage mitigation of industrial pollution. Although national concerns about air pollution have not subsided in the three decades since MacMillan wrote his dissertation, the means to keep sulfur dioxide emissions under control have changed. Strict standards such as those in the EPA's "New Source Performance Standards of 1971" have given way to openly politicized agreements. The Clean Air Act Amendments of 1990, for example, allow plant owners to buy emissions allowances—the right to add sulfur dioxide to the atmosphere from other industrial areas where fewer pollutants are emitted—so they can burn coal with higher sulfur content. The purchase of emissions allowances is a hedge and a subtle support for mining regions that are stuck with moderate to high sulfur content coal. In many ways, this amendment to Clean Air legislation is in keeping with the agreements MacMillan described between the Company and the federal government in the early 1920s. Environmental protection is still accorded second place to continued economic growth and corporate stability.[16]

Donald MacMillan's pioneering work on the environmental consequences of Montana's nonferrous mining and smelting industry is still important after nearly three decades. As an early foray into environmental history, MacMillan's study clearly met the academic standards of the early 1970s, but his book should also be seen in the light of Montana historiography. Clearly influenced by the anticolonialist and anticorporatist tradition that drew a hard line between malefactors and victims in Montana's political and social history, MacMillan edged toward a broader viewpoint on the relationships between industry, government, and political interest groups. By the late 1970s, the Toole and Howard perspective had lost some of its persuasiveness, partly because of the political agreements that the environmentalist and reformist legislators had achieved in their halcyon days between 1969 and 1975. Michael P. Malone and Richard B. Roeder's *Montana: A History of Two Centuries* in 1976 and Malone's *Battle for Butte* in 1981 told the state's history and the story of the great industrial and financial contest for control of the "Greatest Hill on Earth" in terms that blurred the hard lines of the populist historical tradition.

The stories of Butte and mining in Montana history are still being revised. Jerry Calvert's *The Gibraltar* (1988), David Emmons's *The Butte Irish* (1989), and Mary Murphy's *Mining Cultures* (1997) are three superb examples of discussions about Butte's continuing fascination for historians and readers, and also for the need to abandon old and socially binary views of Montana's past. In Butte and elsewhere in Montana, change did not occur on just one or another side of any protracted dispute. History does not work that way. The past is much more gray than it is black and white, and it is a mark of modern historiography about Montana that historians have embraced multifaceted portraits of its past.[17]

Donald MacMillan's book has its feet in both historiographical eras. The legitimate descendant of the Toole tradition of populist interpretations of Montana's past, MacMillan's study of the Anaconda smoke cases is also the beginning of a broader viewpoint on the relationships between the developers of the state's natural resources and the people of Montana. The issues raised in MacMillan's narrative are still relevant and find their way into discussions throughout the American West. In Montana, the advent of the ARCO Superfund cases, the sharp change from a natural resource economy to a service economy, and the continuing interest in environmental issues makes this book important. Those interested in Montana history should read this book as a chapter in a much longer and larger story of the state, a story that includes the air quality hearing in 1972, the fate of toxic wastes in Butte, and the future of Montana's landscape.

– Chapter One –

A STRUGGLE FOR HEALTH, PROPERTY, AND CONSERVATION

THE STRUGGLE TO ABATE AIR POLLUTION in the copper smelters of the Far West between the 1880s and the 1930s took place in three phases. The first phase was an urban struggle waged between the irate citizenry of Butte, Montana, and the owners of the local smelters. In time, because of the increased production of the Butte mines, some of the smelter owners moved their smelters to other locations. When the world's largest and most modern smelter of its time was constructed in the Deer Lodge valley in 1902 to smelt the Butte ores, the second phase of the struggle was initiated. As a result of the relocation, a major portion of the "smoke problem" that had been plaguing the city of Butte was transferred to the Deer Lodge valley. Eventually a bitter contest over air pollution arose between the smelting and agricultural interests of the valley. Ultimately this conflict brought about the third phase, which developed into a classic confrontation between the federal government and one of the most powerful trusts in the mining and smelting industry.

In the first phase, the battle was waged principally in the interest of public health and well-being. There is no evidence to suggest that comparable campaigns to abate smelter emissions were mounted in other communities, but in Butte support for control of air pollution was both broad and deep. In the second phase, the fight was waged chiefly in the interest of property. And in the third phase, the struggle was fought in the interest of conservation.

There were many arguments advanced during each phase of the struggle, but one in particular was consistently advocated, successfully in virtually every case, as the most compelling reason for not effectively abating the poisonous substances of arsenic and sulphur oxides: It was not considered feasible by the smelter owners to eliminate these elements because they were not of sufficient commercial value. Consequently, the experience of these years corroborates what economists maintain: Pollution abatement is not something that can be achieved by the traditional operations of the marketplace. Professor Samuel B. Chase of the University of Montana, drawing on the work of other economists and on engineering studies, pointed out in 1970 that 95 percent of pollution could be "wiped out" within five years if industries would make the necessary investment. Although this appeared to be a bargain, "it is one that the free market, left alone, would pass up."[1]

The focus here is on Montana. Other states, particularly California and Utah, were tangentially involved in the struggle, but no attempts were made to abate air pollution in Arizona, where copper became the leading mineral in the late nineteenth century and provided the base for a stable mining industry. The same is true of New Mexico. It is probable that conflicts did not develop in these areas because emissions did not affect agricultural property or national forests. Only in Montana—and there, only in Butte—was health the paramount concern.

Montana is center stage because it was there that the largest nonferrous mining and smelting trust, the Amalgamated Copper Company, conducted most of its operations and where simultaneously the greatest damage was being done to the national forest and agricultural interests. The federal government came into the case specifically in order to protect the national forest and water resources from the emissions of the Washoe smelter at Anaconda. Public health, while acknowledged as a matter of concern, was subordinate to conservation of natural resources. Most importantly, the federal government concluded that if it could force the Amalgamated's Washoe smelter to adopt an effective abatement program then other smelters in the West would have to follow suit.

The question may be asked why the same intense antagonisms and conflicts that arose in the western states did not arise in cities such as Omaha, Chicago, and New York and in other areas where smelters discharged smoke. Part of the explanation, but certainly not all of it, lies in the reason given by the Bureau of Mines in 1915. Referring to suits filed by agricultural interests and certain other landowning associations in the West during the preceding decade, the bureau said that "none of the metallurgical plants near these cities treats any considerable tonnage of high-sulphur ores. Most of the smelters so situated," the bureau claimed, "are refining plants, which may smelt some sulphide ores, but not in sufficient amount to cause trouble from sulphur dioxide."[2]

The struggle in Montana was conducted with the bitterness and gamesmanship generally reserved to military campaigns. For this reason, the lessons of the struggle, both explicit and implicit, are germane today. The current conflict over air pollution in the smelting industry is not a mutation springing from an unknown parent. It is not a case of society suddenly changing the rules. The present conflict is the legacy of these earlier battles.

The lessons of the earlier struggle might be less compelling if the antagonists in the final phase were not the giants of the age. The battle brought the federal government into the field against the Amalgamated Copper Company. The Amalgamated was actually a holding company controlled by Standard Oil in which the First National City Bank of

New York held a large block of the stock. In 1899, the Amalgamated purchased the Anaconda Copper Mining Company which became its operating company.[3]

Initially, Pres. Theodore Roosevelt, vigorously assisted by his attorney general, Charles J. Bonaparte, provided the leadership for the government's effort. The struggle then passed in 1909 to the William Howard Taft administration to be placed under the capable direction of Attorney General George W. Wickersham.

When it joined battle with the federal government, the Amalgamated Copper Company was the nation's largest and wealthiest corporation in the nonferrous metals industry. Theodore Roosevelt was not only an activist and a powerful president, he was also the country's first effective conservationist. To him, conservation—meaning the preservation, protection, and organized use of the nation's resources—constituted the fundamental answer to almost every problem of national life. He gave the highest priority to what he clearly recognized as a fight that could go far toward establishing the right of government to intervene in the private sector when and where the nation was put upon by corporate cupidity in the form of resource exploitation.

In some respects and at various times, Theodore Roosevelt was an impulsive man. In this engagement he was not. On his personal orders the most meticulous and comprehensive preparations were made for the battle with the industrial giant. The finest legal talent was assembled; scientists and experts were gathered and sent out into the field. Roosevelt and his attorney general studied political repercussions and tactics with as much care as the teams of chemists and botanists studied the effects of arsenic and sulphur oxides on vegetation and animals.

The Amalgamated was not merely a corporate giant; its legal staff was seasoned and skilled. The trust could make or break the careers of thousands of chemists, engineers, and metallurgists; it had tremendous political muscle; and it had the support of the prevailing economic theory of the nineteenth century. Though the economic winds were changing, Amalgamated also had the inestimable advantage of a court system loath to alter the status quo, and even more loath to tamper with the venerated idea of the free market. Not all of the courts were of like mind; not all would fall automatically into line on the basis of the prevailing economic-social theory. But neither Roosevelt's men nor Taft's could count on precedent in that regard. Nevertheless, the battle with the trust was fought because they considered the objective vital to the future well-being of the nation. At the same time they prepared to move against other smelters in the West to force them to abate poisonous emissions.

Eventually a truce was reached with the trust, objectives were formulated, and promises were made that seemed to provide a formula for resolution of the air pollution problem within the smelting industry. Thereafter, similar agreements were reached with the lesser firms. But with the pressure off, the spirit of the prototypical agreement was ignored and, as the rest of the industry looked on and as part of the federal government assisted, the Amalgamated systematically put into operation its own plan for remedy of smelter fumes. Finally, as a consequence of the strategy of boring from within during the deferential administrations spanning the twenties, the work of the Roosevelt and Taft administrations was almost completely undone.

No reasonable person expects the environment to be as pristine as a snowflake. But the regrettable fact is that in the period between the 1880s and the 1930s the people with the most talent and resources—some of them the most respected men in American society—were the least disposed to preserve and protect the nation's natural resources, despite the strenuous efforts of reformers. At bottom, this struggle with the smelters was an effort to balance public interests and private. To achieve that balance, coercive force had to be exercised on a national scale by a responsible public. The only agency with the resources capable of establishing and maintaining such a public policy was the national government. It was this fact that Theodore Roosevelt recognized in his battles with the private interests.

Until Roosevelt, the attempts to abate air pollution had been admirably articulated, but they lacked the wherewithal to contend with the power of the smelter owners. Looking back upon his administration, Roosevelt observed that, before his presidency, "private rights had almost uniformly been allowed to overbalance public rights. The change we made," he concluded, "was right, and was vitally necessary."[4]

It would make for a neat monograph if one could confine this story to that of a simple clash of the titans. But the clash itself has to be put in context. So the story must begin where it began—in the mid-1880s in the roaring, rumbling city of Butte, Montana Territory. And it begins with two men, one a local cattleman and the other a friend visiting from England. Stepping off the train from Chicago one late afternoon in December 1885, the two men carefully made their way to Butte's small depot, its lights barely discernible through the volumes of yellowish smoke thick with the fumes of arsenic and sulphur. It was "like a pall enveloping everything in midnight darkness and almost suffocating," the cattleman, Granville Stuart, wrote later. "We could not see and we could scarcely breathe. The Marquis grabbed my arm and between sneezes gasped—'What is this to which you have brought me?'"[5]

– Chapter Two –

"The Humblest Citizen of Butte is Entitled at Least to Fresh Air"

IN THAT YEAR OF 1885 BUTTE lay sprawled on the side of a mountain at the northern extremity of Summit Valley just below the Continental Divide. The valley resembled a bowl, rimmed on three sides by the main range of the Rocky Mountains. Lesser hills lay to the west. Several small communities surrounded the city in haphazard fashion, extending above it to the mountain's summit and below it onto the valley floor. During the winter months, and occasionally in the summer, an inversion often placed a lid over the valley and trapped the city's smoke and airborne debris halfway up the mountainsides.

Beneath the city lay one of the richest mineral deposits in the world. Butte was born a gold-mining camp in the 1860s; by the 1870s it was a roaring silver camp, and by the time the cattleman and the marquis arrived, it had become a respectable copper producer, attracting capital investment. By that time, it was shedding the ephemeral and informal characteristics of a camp and adopting the permanent, if raw, institutions of a frontier city. Energy, industry, and high hopes permeated the region. The population, mostly immigrant, numbered approximately six thousand, with perhaps another five thousand in the satellite settlements.

There were more than three hundred mines, nine stamp mills to crush the quartz ores, and as many as seven copper smelters operating within and around the city. The ceaseless rumbling and falling of the stamps in the mills shook the smoke- and dust-ladened air. If a breeze or wind was not up from the north or west, the smoke and particulates settled on the townspeople—burning eyes, searing nostrils, and clogging throats.[1]

The increasing pollution and its devastating effects prompted the first public protests. It also prompted reminiscences among some of the old-timers about how serene and pristine the countryside had been before the advent of the smelters. Jailer James Cook, one of Butte's earliest residents, recalled how green the grass was when he came, how it grew so thick and long that cattle were often lost in it. A contemporary chronicler wrote of how evergreens covered "the gently undulating foothills" that circled the town. Vegetation and a rich array of wildflowers covered the valley floor. Strawberry and alder bushes grew in abundance along the creek that wandered through the valley. A group of ladies from the city's west side, concerned with "beautifying their homes," initiated a heated protest to city officials and the newspaper about the "clouds of sulphur smoke which hung over the city Sunday night" and killed most of their flowers and plants. Their efforts were rewarded with condescending humor—and neglect.[2]

If the gentle ladies of the West Side Garden Circle and other aesthetes mourned the area as a botanical wasteland, few of their neighbors at that time shared their sentiment, even though they too suffered the discomfort of the smoke and fumes. The people who came to Butte in the early 1880s were seeking not a pastoral paradise but the possibility of advancement. To the migrating laborer and immigrant, with all-too-recent memories of deprivation and hardship, advancement meant steady work and perhaps even promotion. To the experienced entrepreneur out of the East, it meant eventual wealth and social status. The comment one of the earliest and toughest of these eastern enterprisers made to a colleague is at once representative and to the point: "I want to make big money out of copper," he said.[3] The din of the stamp mills and the smoke from the smelters were to most residents genuine evidence of an advanced civilization.

The collective attitude about the environment and the future of the region in the mid-1870s to mid-1880s was probably best expressed in the centennial speeches of two of the city's most prominent citizens: William Andrews Clark, banker and mine owner, and Gen. Charles S. Warren, Civil and Indian War veteran and former mine owner. Smoke gave meaning to the lives of these men. It evidenced the "busy hum of industry." The mingling of the smoke with "the pure air that envelops her mountains," they told the admiring crowd, testified to Butte's advent as "the future center of trade, industry, and wealth of the great northwest" and "the great mining center of the West, if not the world." Soon capitalists throughout the world would seek investment and "the artisan and miner would have constant and profitable employment." Clearly, a new day had dawned for all.[4]

In the last decade and a half of the nineteenth century, Butte mirrored the nation's incredibly rapid industrial expansion and economic concentration. Within five years the town's copper production almost doubled, increasing from nearly 68 million concentrated pounds in 1885 to almost 113 million in 1890.[5] In the same period the mines were consolidated and an interlocking directorate was formally introduced. These two developments meant that control of the region and its resources went to outsiders, all of whom shared a common interest in the maximization of production and profit and a common indifference to the community's well-being.[6]

Within three years, between 1887 and 1890, most of the smelters had increased their productive capacity twofold. The Butte and Boston Mining Company, as an example, increased its output of fine copper

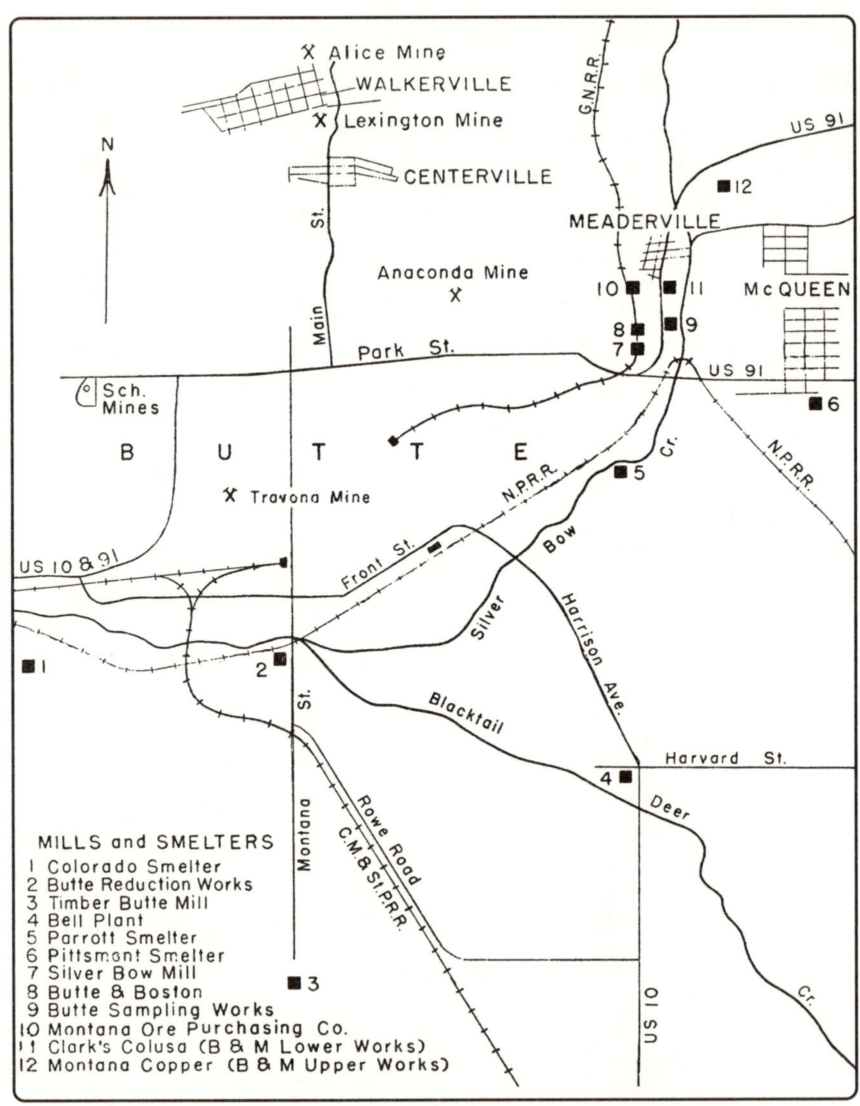

Map of Butte, showing the locations of early reduction plants.
RALPH I. SMITH, *HISTORY OF THE EARLY REDUCTION PLANTS OF BUTTE, MONTANA*
(REPRINT FROM *DE RE METALLICA*, 18 (MAY 1953), 1-19)

within those three years from less than 1 million tons of fine copper per month to 2 million tons per month.[7] In 1890, six of the most modern and potent copper smelters in the world were strung out in picket-line fashion along the base of Butte Hill, filling the valley with so much smoke that one visitor, touched by the sight, wrote to his son, "This is a big town, but between the smoke from the smelters and the burning timber it is not easily seen."[8]

What the visitor referred to as "burning timber" was open-air roasting of ore, a process called "heap roasting." It and a variant, "stall roasting," were the first phase of the smelting process and the principal source of the suffocating smoke. With the tremendous increase in the production of the mines, the critical problem was how to refine the ore locally in order to obtain a higher financial return by reducing the freight weight transported out of the region. Roasting the ore in heaps was conveniently cheap and reasonably efficient. Large lumps of almost pure sulphide ore were intermixed with layers of logs. The size of the heaps varied, some being as long as a city block, as wide as a city street, and perhaps six feet in height. Ignited, the heaps underwent a slow burning process that removed all but 8 percent of the sulphur from the ore. For as long as two or three weeks they burned and smoldered, continuously releasing clouds of toxic smoke and fumes composed of undiluted oxides of sulphur and arsenic, particulates, and a host of fluorides.[9]

Stall roasting employed the same principle as heap roasting, but the ore was dumped into stalls. Some stalls had stacks, but most of these stacks were too small to be effective. Each smelter used either one or both methods simultaneously while roasting a finer grade of ore in a variety of furnaces that emitted their smoke and fumes through stacks.

In 1891, a representative of the Boston and Montana Company, one of Butte's three biggest providers of ore, wrote in the prestigious *Engineering and Mining Journal* that about twenty-five thousand tons of ore were "constantly in process of open-air roasting," and a "good deal" of the refined and finer ore was roasted in furnaces. Three of the most productive smelters employing these methods—the Butte and Boston and the upper and lower plants of the Boston and Montana—were situated just beyond city limits, directly east of the heart of Butte, in the settlement of Meaderville. When the wind was from the east, or with no wind, clouds of smoke hugged the ground and inundated the town.[10]

If its mineral had made the town famous, its environment now made it an ecological disaster. By 1890 practically all the vegetation in Butte and on the surrounding hillsides had disappeared. For the four trees that survived in the city, there was, one newspaper reported, "general commiseration." During the course of one inversion, when heap roasting was not being conducted, a journalist climbed a mountain that overlooked the city and recorded his impression. "Last night and today an excellent opportunity was afforded for a study of the smoke question," he wrote. "The atmospheric conditions were just right for complete envelopment of the city in a cloud of smoke and fumes, the air being damp and heavy and almost perfectly calm. All day long the smoke arose from the works below the city and stretched across the valley and against the mountains on either side." The smoke rose no higher than the town, and the reporter could observe a line of smoke extending east and west along the base of Butte Hill. The sight struck him as the "shores of a still, calm sea of deep murky water, which buried everything below it." Occasionally, he recorded, "the sound of traffic from the busy city hidden in the smoke" drifted up the mountainside. "Every prominent building and landmark was obscured and only the surface of the silent mysterious sea below could be discovered." From time to time "a street car could be seen emerging from the line which marked the meeting point of the smoke and land." It "looked uncanny and strange, appearing to suddenly rise from an immense body of water."[11]

Inversions could last for days, and during such times the citizens literally groped their way around town. Carriages had to be driven slowly for fear of knocking pedestrians down. Railroad engines collided with one another in the switching yards. Trolley cars had to creep through the streets, ringing their bells constantly; at night the conductors walked ahead of the cars with lanterns in their hands. Many people experienced bleeding from their noses and some vomited in the streets. Workers would lose their way going to and coming from work.[12]

Some vocalists on nationwide tours were in constant dread of losing their voices to the smoke while in Butte. Emma Abbot, a prominent singer of the day, was so frightened when a tour brought her through the town in 1890 that she had heavy blankets tacked over all the windows and transoms and "stuffed up the keyholes" in her room at the fashionable McDermott Hotel. Fortunately for the city's music lovers, an admiring reporter told the town, Miss Abbot's "sweet notes . . . lost not a whit of their sweetness through Butte's smoke."[13]

In a debate about the comparative merits of the territory's communities at Montana's 1889 constitutional convention, one imaginative delegate recounted his short stay in the mining town: "I had not been in bed thirty minutes before I felt that some demoniacal fiend was there injecting boiling lime into my lungs." For the citizens of Butte, he concluded, "it must be very consoling to know that the transition from this world to Sheol will be so slight."[14]

The smoke was not without its defenders. Some thought it benign and salutary. Most prominent among them was the aforementioned William A. Clark, by now owner of numerous mines and a large smelter in the valley. He was also an aspiring magnate and hungering after national political office. In an address before the constitutional convention, Clark endeavored to rebut the testimony of the delegate who had suffered at the hands of the "demoniacal fiend," admitting that there may have been times when smoke settled over the city but that Butte boasted a "refined and elevated society." Actually, the smoke was a disinfectant that destroyed "the microbes . . . [that] constituted the germs of disease," Clark claimed. Although "disagreeable in some respects," he said, "it would be a great advantage for other cities . . . to have a little more smoke and business activity and less disease." Not only was smoke good for business and health, but it was a treasured cosmetic as well. Revealing what must surely have been a well-kept secret, Clark told the assembly, "I must say that the ladies are very fond of this smoky city . . . because there is just enough arsenic there to give them a beautiful complexion, and that is the reason the ladies of Butte are renowned wherever they go for their beautiful complexions."[15] If few residents had agreed with the ladies of the garden circle in 1885, even fewer agreed with the august mining baron some four years later.

From July to October 1890, there were 192 deaths in Butte, most as a result of pneumonia and typhoid, more deaths than residents could recall having occurred before.[16] The smoke was proving to be a scourge rather than a harbinger of wealth, security, and progress.

Then in the winter of 1890–91 the city experienced the worst period in memory of what had come to be called the "smoke season." In the month of December smoke covered the city twenty-eight times. The *Anaconda Standard* asserted that "the mortality of Butte [in] the present month is frightful." Of the total number of deaths that month, thirty-six were ascribed to breathing-related diseases. "One of the best physicians in Butte said today," the *Standard* told its readers, "that it is almost impossible to cure either pneumonia or typhoid fever without pure air. For recovery the lungs must have pure air, and the smoke is of course suffocating. It is the opinion of good physicians that but for the smoke several, if not most of those sick with pneumonia would have recovered."[17]

Butte, Montana, August 1890. Photographed by F. Jay Haynes.
Haynes Foundation Collection, Montana Historical Society Photo Archives, Helena

Few occupations were spared. Of those listed among the dead were a prostitute, a gambler, an actor, a traveling salesman, and a number of laborers. The greatest mortality, of course, was among miners. They were especially vulnerable to pneumonia and tuberculosis as a result of the irritation of the lungs from silica dust in the mines, and then after working long, arduous hours in the hot, dark tunnels, they emerged abruptly into the wintry weather for a chilling walk home through poisonous smoke. A deep malaise spread throughout the community.[18]

Especially disconcerting was the description of the death of one citizen headlined in the *Standard*: "Garing's Strange Death: While Standing By His Bedside He Quivered And Died." Fred J. Garing was a miner "only about thirty years old and prior to the strange malady . . . was considered the embodiment of manly vigor and strength. He was of a quiet disposition, frugal and industrious and had amassed considerable property." Garing owned his house "and had quite a little sum of money in the bank." He had worked his usual Wednesday night shift "and made no complaint of feeling ill until he went off his shift, when he said he had a pain in his throat." The next day he was examined by Dr. Thomas Murray, who found "him suffering from congestion in the throat, apparently caused from a bad cold, and prescribed accordingly." Experiencing no relief, Garing returned to the physician, who reaffirmed the original diagnosis but gave him a different prescription. "By the time he arrived at his room he rapidly grew much worse and his throat so nearly closed up that he was unable to take any medicine, and while standing by his bed fell dead to the floor without a word."[19]

In September 1890, over the objections of city officials concerned about the town's image, the city health officer had insisted on publishing the monthly death rate. At year's end, he computed that for a population of thirty thousand in Butte and vicinity the death rate that year exceeded the 1889 death rates of the highly industrialized and urbanized cities of Chicago and London and almost equaled those of New York and Paris.[20] The city had been buoyed by the earlier prophets' vision of a great and prosperous commercial center, but chronic sickness and high death rates had not been in the reckoning. Pollution had come too far too fast. Consequently, a movement to abate the smoke was quietly launched by some of Butte's citizens.

The leadership came from a crusading physician and a minister. Dr. Heber Robarts and the Reverend Jerry Rounder were archetypes of a new element being introduced into American society by reformed medical schools and the Protestant "social gospel" movement. These young men rejected the callous materialism of the age and chose to dedicate their lives to careers of service. Their ideals were clearly out of keeping with the prevailing national dogmas of laissez-faire economics and social Darwinism, but a generation coming would embrace their doctrines and carry many of them to fruition. Dr. Robarts and Reverend Rounder reasoned that the public had the right and duty to regulate business enterprise in the general interest of the community and in the specific interest of its less fortunate and more vulnerable members. In the words of the historian Howard Mumford Jones, they were part of the magnificent "race of dreamers and reformers . . . the like of which we had not seen before."[21]

Reverend Rounder was one of a small minority of Protestant clergymen who turned away from the traditional emphasis on spiritual and moral concerns. Their gospel stressed a human's relationship to other humans rather than to God. They called for social reform and betterment. Where the more respectable clergymen in town preached about "the forces of darkness," Reverend Rounder preached about the people and forces responsible for the darkness over Butte. In the early period of the antismoke campaign, much of the citizenry, especially the influential, dismissed his sermons as "the usual Sunday evening harangue," but as the smoke and fumes grew deadlier, Rounder's sermons suddenly became "eloquent and common-sense discourses," as he flayed the community for its apathy toward existing conditions.[22]

Reverend Rounder saw the source of the morbid environment in the values of society. One Sunday, with complete disregard for popular prejudice, he directed the citizens to emulate what many considered a "lesser breed" if they expected to live a decent life. The title of his sermon that evening was an effrontery in itself: "Go to the Chinaman: Consider His Ways and Live." Upbraiding the citizenry for the blatant racism and discrimination shown Butte's large Chinese population, Rounder said:

> If the people who are worrying about the mortality of Butte would study the Chinaman's ways and pattern after him they would do a great deal better, and there would have to be less hard ciphering in explaining away those health statistics. . . . He doesn't work all day,

> make the rounds at night, playing faro and bumming, and swilling whiskey to float a ship and then work all the next day. . . . He isn't running for mayor and he isn't in the hands of his friends! . . . The Chinamen, furthermore, has [sic] no expectation of getting rich in a minute and a half. They don't worry over becoming millionaires immediately. They take life philosophically and they are living while we poor civilized inhabitants of Butte are dying.[23]

Never letting up, Rounder queried his flock another time, "Do you look upon Butte as a growing city, or merely as the greatest mining camp on earth? Do you regard smoke as a healthful, God-given blessing? Do you consider it a sin to do all you can toward abolishing the sulphur smoke?"[24]

Rounder's cohort, Dr. Robarts, was a practicing physician who had been appointed city health officer and head of the rudimentary health department in 1889. Prior to his appointment, the health department was little more than an inefficient board. Much to the distress of city officials and many of the citizens, Robarts took his position seriously. Urbane and understanding, he was a tireless missionary for public health, continually calling for improvement of the community's environment and its cultural life. For the public's health, preventive medicine was the goal. "To guard is better than to heal. The shield is nobler than the spear," he admonished city officials. At a time when Americans measured their civilization by a narrow, quantitative ethic, Robarts appraised it in qualitative terms:

> The real wealth of the people is not counted by its gold, silver and acres. These are sources of material and physical greatness. Above these . . . is the health of the people. Here is the manhood, the real civilization, the source of its content, happiness and its good will to man. . . . Life and health to all is sacred. . . . A beautiful city is the mark of civilization, stately school houses and gorgeous churches mark the tread of men of letters. Yet of these we have but little. . . . Man should reach with little suffering his three score years and ten.[25]

Robarts's humane world was full of promise. Not appreciating the correlation between industrialization and an increasing cancer rate, he predicted that "Cancer shall meet its enemy within the vision of our time." He pressed for a citywide program to eliminate the "preventable diseases." But of all the morbidity afflicting the community, his major concern was the "evil of the smoke."[26]

Dr. Robarts concluded that there was a direct relationship between the smoke and fumes of the smelters and death and disease in the community, something which, though continually alleged, was not demonstrated until the late 1960s, when research showed that air pollution can cause the defense mechanisms of the pulmonary system to break down.[27]

Dr. Robarts argued his case against pollution before the Butte City Council at every opportunity: "Those who say the sulphur smoke is healthy talk nonsense," he told city fathers. "The facts are that there is not a solitary disease germ that is affected by the sulphur fumes of this town. . . . It has been apparent that when the smoke rolls into the town with a great deal of sulphur and for long duration the return of death certificates are more numerous, especially pneumonia."[28] A few months later, he reported, "The sulphur is a direct, painful irritant to the glottis and respiratory tract. The oxygen of the air is replaced by sulphur. . . . In cases of pneumonia it greatly increases the danger [of death]. In pneumonia and typhoid fever it is necessary to have fresh air."[29]

Although idealistic as to objectives, Robarts was realistic as to means. He articulated his abatement program in language and in objectives to which the public could relate. The basic reason for controlling the smelter smoke was to achieve the broad goals of a community. Smoke "is the most mortal thing to the individual . . . and the thing most detrimental to the further growth of the city," he told the city council. "Good health statistics will afford a good idea of the measure of progress a community has made in civilization." Statistics are "a valuable guide to the legislator." They "indicate growth and prosperity, the places of good investment," Robarts said. He knew who suffered most and anticipated who would bear most of the costs of abatement. "We owe the abatement of this nuisance to the poor people of the city. It is not chiefly for the benefit of the rich men. They can move two or three miles out and come into the city for business alone, but the poor man cannot." The poor were "also the men who would put up the money," Robarts claimed.[30]

In November 1890 the efforts of the two reformers paid off when the *Anaconda Standard*, the daily with the largest circulation in town, joined their campaign. Though located in Anaconda, the *Standard* was essentially a Butte paper. Its planned circulation zone was "Butte-Anaconda," and Butte accounted for the greatest part of its circulation.[31] Once committed, the paper made the antismoke campaign a crusade and eventually moved to the vanguard of its leadership. Following the *Standard's* lead, the *Daily Inter Mountain*, the other daily in the city, gave its support to the movement. Pointing out that the smelting companies were "all rich and making plenty of money," the *Inter Mountain* editorialized that "The humblest citizen of Butte is entitled at least to fresh air and is justified in objecting to a continuance of sulphur and arsenic which though not unhealthy, is deucedly uncomfortable."[32]

Earlier in the month the *Standard* had run a story about how the smoke prevented a "boom" in Butte real estate. Real estate was "far below what it is in other cities the size of Butte," the paper explained. But, it continued, "Health Officer Heber Robarts said today that Butte's smoke can be removed and he has a scheme on foot to bring this about. . . . No finer place could be found to live in with the smoke gone, grass and trees would come, and people who cannot live here now would prefer Butte to any city in the country. . . . It would be worth hundreds of dollars to every citizen of this city to get rid of the smoke, and the thing can be done."[33]

Once the cause of a clean environment was joined with the dollar, the movement gained momentum. The board of trade, an organization of local business and professional leaders, comparable to a present-day chamber of commerce, formed a committee, with Dr. Robarts as its director, to investigate methods for the "destruction of the gases and smoke emerging from stacks, resulting from burning copper ore," which caused "the evolving of large quantities of sulphur and sulphur gas." The committee wrote letters to the *Scientific American* and the *Engineering and Mining Journal*, asking for a solution to their problem.[34]

A host of replies came from throughout the country. The *Standard* referred to the writers who offered solutions as "candidates for immortality" and as the "smoke Gods of Butte." The writers were generally identified as metallurgists or "chemists of no mean repute." In addition to offering a remedy, they invariably guaranteed a by-product, or as some chose to express it, "another source of wealth." Generally, the solutions were of three types: (1) conversion by combustion of the smoke and fumes into sulphuric acid, arsenic, or lime; (2) removal of the smoke and fumes by crude chemical "scrubbing" devices; and (3) smokestacks two hundred feet high that, having released the emissions at such great height, would allow the smoke to be wafted into the cosmos, never to return.[35]

Such solutions became the "all-absorbing topic" of the city. One of the most novel was proposed by C. W. Stickney, noted mining engineer for Butte's Parrot Silver and Copper Company. Stickney reasoned the "most feasible solution" was to dig a huge trench and channel all the smoke from the smelters into it, then lay a box over the trench level with the ground. Theoretically, the prevailing westward wind would carry the smoke eastward inside the box and release it miles from the city. Not surprisingly, other experts dismissed this as impractical and too expensive. Most thought the conversion of the sulphur smoke into sulphuric acid was the "most practicable." The *Standard* reported that many "seem to consider that the conservatism of the smelter owners is all that stands in the way of consumption in this manner."[36]

If by "conservatism" the *Standard* meant economic feasibility, it was exactly right. Charles H. Palmer, superintendent of the Butte and Boston Mining Company, emphatically told a *Standard* reporter that "There is no doubt in the world that the greater part of the sulphur can be eliminated from the smoke. But the product is sulphuric acid, and the only places where the sulphur has been eliminated are those places where there is use to be found for the sulphuric acid."[37] In short, since there was no local market for it in the Butte region, the only alternative was to dump it over the town and countryside.

The town's most prominent resident, William A. Clark, could contribute little more than fulsome rhetoric to the debate. Through a spokesman he announced that the town's concern was appropriate and assured all citizens that "he was not indifferent to the question of how to get rid of the smoke." He had "devoted much thought to it but could learn of no practical solution," he explained. The proposals he knew of were "impracticable on account of the great cost." Nevertheless, he was willing to "second any practicable movement to get rid of the difficulty" and suggested "invoking the aid of science before undertaking costly experiments." One wit, apparently perplexed by the opposing arguments of the experts and having little confidence in the aid of science, remarked in exasperation: "Why not just grade down the range of mountains behind the town and let the wind carry it away?"[38]

Over a period of weeks, concern for eliminating the smoke had become widespread. As a result of the continuous agitation of the abatement leaders and the *Standard*, the board of trade's committee had been made a component of the city council and designated the "smoke committee." The committee consisted of Dr. Robarts, three council members who were prominent businessmen, and three professionals (two were mining engineers employed by the Anaconda Mining Company and the Parrot smelter, and the other was a chemist employed by a local assaying company).[39]

The first thing the committee did was to tour the smelters and interview the smelter managers with the object of securing their cooperation in abating the smoke. They visited every smelter in the region. Each manager they talked with reiterated Superintendent Palmer's view that converting the sulphur into acid was technologically possible but economically infeasible. "The cost of converting the smoke into sulphuric acid would be very considerable and we could not dispose of it," explained J. E. Gaylord, general manager of the Parrot Silver and Copper Company. "It would cost more," he emphasized, "to ship the stuff out of the territory than to manufacture it outside." He would not agree to closing down during the three worst months of winter unless "an arrangement could be made with the Anaconda company by which they would also close."[40]

In Henry Williams, superintendent of the Colorado Smelting and Mining Company smelter, the committee met with portent of what they could expect from some quarters in the future. According to a *Standard* reporter accompanying the group, Williams felt that "nothing but mischief [would] come" from an attempt to abolish the smoke. There were a number of remedies "but the smelter might as well close up as to use any of them," he said. For the complaints of residents, Superintendent Williams had no sympathy: "The Colorado company was here before most of the residents and those who came in later knew what to expect." Besides, he maintained, the sulphur in the smoke was healthy and "not strong enough to affect animal life." He knew of "a cow which had thrived for fourteen years in the midst of the smoke but had died immediately on being taken out of it." To participate in a planned meeting of the city council to discuss abolishing the smoke seemed, to Henry Williams, "like attending one's own funeral."[41]

In spite of the discouraging attitude of the smelter operators, the smoke committee pushed on and with the *Standard's* help managed to convince most of the citizenry that open-air roasting, considered to be responsible for two-thirds of the smoke, should be abolished. Once that process was abolished, the work could go on to accomplish the complete abatement of the smelter emissions. It would not be easy. Opinion was still divided. In the course of the debate, rumors were floated that, if open-air roasting was abolished, the Boston and Montana Company would have to close down and move out of the region. Some opposed any dramatic action that would jeopardize jobs and trade. Others were undecided about the nature of the smoke; Henry Williams was not alone in maintaining that the smoke was healthful even if disagreeable. Certain people thought it acceptable if either the sulphur or the arsenic was extracted. And still others wanted "the stuff disposed of simply because it [was] smoke and disagreeable."[42]

On December 3, 1890, Robarts proposed the first smoke ordinance to the city council. It called for the abolition of open-air roasting of minerals within three miles of the city limits, except where the emissions passed through chimneys or stacks. Disagreement developed over how high the stacks should be and the ordinance was not passed until two weeks later, on December 17. In its final form, it passed unanimously and differed from the original proposal in that it declared open-air roasting a public nuisance and imposed a fine of from one hundred to three hundred dollars upon conviction. Ores, however, could be roasted where the smoke passed through chimneys or stacks seventy-five feet high. The *Standard* was jubilant and glimpsed the end of the smoke problem. "Public spirited men among the smelters owners" were willing "to adapt their works to the existing conditions. Others less liberally disposed will simply have to fall in line," the paper declared.[43]

On the surface, because of the inconsequential height required for the stacks, the smoke ordinance appeared as nothing more than a Pyrrhic victory. But to the leaders of the abatement movement, the height of the stacks did not matter; the important achievement in their minds, and the reason for the *Standard's* enthusiasm, was that thereafter all emissions were required to be funneled through a stack.[44] The city council had settled the issue, and now it was merely a matter of choosing and installing a proper device in the smokestacks.

Since November the smoke committee had been corresponding with William Hutchinson, an uncommonly persuasive and enterprising inventor from Chicago. Hutchinson was not without credentials. He had something of a reputation for having shown Chicago the way to solve the smoke problems caused by the burning of coal. He had developed a method for converting steam-boiler furnaces into gas furnaces. The *Industrial World*, writing about his work, commented, "we have means at hand to abate the nuisance practically," and "the time is not far distant when to see large volumes of dense black smoke rolling from the chimneys will be considered a disgrace."[45]

In a series of convincing letters, Hutchinson told the smoke committee just what it had wanted to hear: "There is no necessity for any city to suffer from the smoke nuisance, and they are very foolish to stand it," he wrote. He would not divulge the details of his process in writing, but he hinted at having the capacity to destroy arsenic and sulphur completely within the stack by "extremely simple" means and with little expense. "When the people are ready to move, I would be glad to reach them through you and make them estimates and plans that would accomplish the result."[46]

Professionals expressed reservations. The chemist, Prof. Louis Jennin, and the engineer from the Anaconda Company, W. G. Cole, claimed that Hutchinson did not understand the nature of Butte's problem. The problem was removal of sulphur and arsenic from copper ores, not from coal, quite a different thing in their minds. "You can't alter the laws of chemistry. They are just as stable as the laws of gravitation," Professor Jennin argued. But others were convinced. The *Standard*, in its eagerness to correct an abuse, had allowed conviction and hope to supersede evidence and discretion. Several days before Hutchinson arrived, it hailed him as the "smoke messiah" and told its readers that "If the smelters will not accept his process they can be forced to by law."[47]

The "messiah" reached Butte in mid-December, at the height of the debate over the smoke ordinance. Through his demeanor and unbounded confidence, he captured the city's imagination. Soon he was being called "Old Hutch." Old Hutch did not "deal in SO_2 and SO_3 and other generalities and theories"; he was a "thoroughly practical man," and all business, said the *Standard*. He had an "inborn antagonism to smoke," it was said, and he always kept an unlit cigar in his mouth, biting and chewing at it continuously, thereby showing his disdain for smoke.[48]

At the end of his first day in town, after a whirlwind tour of the smelters, Hutchinson held an "after dinner interview" for the gentlemen of the press at his hotel, the McDermott. Talking around his cigar, he announced, "I can state positively that I can get rid of your smoke. There is no doubt about that at all. I have seen nothing new as yet. . . . I have studied smoke for twenty years in a good many different cities. . . . I tell you that sulphur can be burned in any form or condition. That is a fact that is denied, but is nevertheless a fact." He referred to this method variously as "fire boxes" or "smoke consumers" that could be installed in the smelters at little expense.[49]

Old Hutch allowed the town no doubt that he was their deliverer. "The people must not have an idea that I can come here and with a simple sweep of the hand drive away the smoke," he said. "If the smoke is to be gotten rid of it must be gone about in a businesslike manner. The people must demand it. It must be a popular movement or it will never succeed." He told the citizens of Butte exactly what they could expect:

> I am not a philanthropist. I do not go around as a public benefactor. I am a businessman pure and simple. I am in this thing to make money, not for my health. I have made a good deal of money in various cities by my smoke consumers, and I can make money here if the smelters will give me a chance. I must have the cooperation and support of the sweltermen. I am now ready to contract with the sweltermen to do away with their smoke. If they want to get rid of it that is all the difficulty in the way. I will not charge a solitary cent unless I succeed. . . . If I don't succeed in getting rid of the smoke I alone will be the loser.[50]

Although Hutchinson promised "exact terms," roughly calculated at two hundred to four hundred dollars per furnace, contracts from the smeltermen were not forthcoming. Therefore, on the day before the smoke ordinance was passed by the city council, the smoke committee arranged an agreement whereby Hutchinson would return to Butte within a month to "destroy the smoke of the Parrot plant." On the day of his departure, Old Hutch felt constrained to assure the town and, presumably, disparage his critics by reiterating to newsmen: "I want to say again and finally that sulphur can be burned in every form and condition. . . . I know what I can do. I know just what I have to deal with."[51]

These had been heady days for the progressive element of the city. It had initiated a reform movement; it had rendered the smoke, formerly a symbol of achievement and beneficence, a mark of excess and morbidity; it had passed a remedial ordinance and hit upon a wondrous device that promised relief. Dr. Robarts expressed the reformers' optimism in a talk before the city council after the ordinance had passed and Hutchinson had left the city: "The time has come when the masses must be aroused to a more earnest sense of the people's welfare and adopt the means that will protect us from disease and better develop the physical, moral and intellectual powers."[52]

The future was bright. Butte was patiently awaiting the return of the "smoke messiah," said the *Standard*, "in the belief that by the magic of his word and his patent[ed] machinery the plague of smoke will be removed forever."[53]

– Chapter Three –

"Bluffs Don't Go in Smoke Wars"

William Hutchinson, the smoke messiah, returned to Butte in the middle of January 1891, when the city was undergoing one of its worst experiences of smoke and fumes. Dr. Heber Robarts had initiated the practice of taking weather readings three times daily at various times throughout the day. The presence or absence of smoke in the city was part of the report. Out of ninety-three observations in January, smoke covered the city forty times, twelve more times than in the previous month. During one eleven-day period, smoke filled the city every day. For eight consecutive days during that period, smoke and fumes had engulfed the town at all three observations. In the midst of all of this, the *Standard* masked its growing concern by remarking whimsically, "Pneumonia is becoming quite prevalent again in the existing smoky atmosphere. . . . There was smoke in Butte, and bad smoke, too, every day. The mercury in the thermometer got tired of breathing smelter smoke and took a tumble into the zeroes to escape it."[1]

With his return on January 20, Hutchinson became the center of attention, and his activities were attended with great fanfare. As the town's misery grew, so too did Old Hutch's image as messiah. Unfortunately, his first day at work was not an auspicious one. He had instructed his assistants to be on the job at the Parrot smelter before dawn. They had dutifully left their hotel at six that morning but had gotten lost in the smoke and ended up wandering about the southern part of town until noon, when a wind lifted the smoke. Old Hutch himself fared little better. He had awakened early enough, but "it was so dark from the smoke he didn't know it was morning yet and overslept." He found his way to the smelter shortly after his workers.[2]

It was never exactly clear just what the end product of Hutchinson's process would be, though at one time he hinted to the health officer, "Water is noncombustible unless raised to such a heat as to burst it into its elements. So sulphur dioxide is non-combustible, but when raised to such a heat as to burst it into its element, it too burns." What little he would reveal directly suggested there would be no by-product requiring expensive disposal. During the course of installing his "fire boxes," he explained his theory and confirmed his supporters' brightest hopes. The steam and air "mingling with the sulphur fumes accomplishes combustion of the latter and carries them thru the stack to the air where the residue evaporates." Dr. Robarts, with as good a knowledge of chemistry as most at the time, thought the process was "the evolution of a genius" and "one advance step in chemistry."[3]

On January 24, preparations for Old Hutch's demonstration were completed. The town could scarcely contain itself. "Never before in Butte's history were the prospects so bright for getting rid of the nuisance," said the *Standard*. "Seldom has there been more excitement. . . . Real estate took a decided leap upward this morning as soon as the morning papers were circulated." By one o'clock a large crowd had assembled around the furnace at the Parrot smelter. Included in the assembly were the mayor and councilmen, representatives of the press, and other assorted local dignitaries.[4]

For the demonstration, two furnaces had been disconnected from the smelter's stack and connected to a furnace containing Hutchinson's apparatus. The smoke from the furnaces entered the experimental furnace directly below Hutchinson's firebox. Six pipes, one to one and a half inches in diameter, forced air and jets of steam into the firebox chamber. The smoke at the bottom of the furnace was drafted through the intense fire in the chamber of the firebox and out the stack. A brick below the firebox had been removed so curious spectators could peer into the bottom of the furnace at the "dense cloud of smoke." And above the firebox a steel door was opened so they could witness the transmutation. "We are ready," declared the plucky inventor, "we are ready to turn in the smoke. If the smoke doesn't burn I'll go right uptown, settle my hotel bill, and leave for the East on this evening's train."[5]

The following day the *Standard* told its readers what it had beheld. "Mr. Hutchinson's Condenser Works to the Queen's Taste," it headlined. The fire was "500 degrees hotter than any other fire in the smelter. . . . Above the fire not a particle of smoke can be seen. Blue flames occasionally appear, showing to all appearances the burning of the sulphur. The smoke is dense below the fire; it must pass into the fire, and it cannot be seen above the fire." As far as the *Standard* was concerned, the town had been delivered from the plague.[6]

Others were not so sure. James Breen, an official of the smelter, contended that while the smoke was invisible because of the intense heat, it reappeared in the form of smoke as soon as it struck the outer air. He had watched the top of the stack and noted that for a distance of several inches there was no smoke, but further up a dense cloud had formed. The mayor, Henry G. Valiton, however, termed it a "grand success." "The smoke certainly disappears," he said. "It seems to me we are on the right track." Others were of a similar view yet cautious. Alderman John F. Cowan said reflectively, "I don't know whether the smoke burns or not. It disappears; that is certain. Whether it subsequently reappears in another form, I don't know. But I believe that we are far nearer the solution of the great problem than we ever were before, and that every possible opportunity should be given for a thorough test of the apparatus."[7]

A more thorough test was arranged for "the doubting Thomases," with the city standing the expense. This time the Parrot disconnected two furnaces from the stack and erected a temporary smokestack. Hutchinson's firebox was placed between the furnaces and the stack. "If no smoke appears above the stack that will be proof indisputable," the *Standard* wrote. "If smoke shows in any large quantity, that will show that the smoke, after being burned, is again condensed into smoke."[8]

In the meantime, members of the city council had pressed for an ordinance that would force the smelters to install Hutchinson's device, provided the new test proved satisfactory. Echoing the council's sentiment, the *Standard* asserted, "Even if there are sulphur smells, but no smoke, that will be sufficient for the people. To get rid of the sight of the smoke will be a marvel."[9]

The second test was anticipated with even greater excitement and attended by a larger crowd. The results were essentially the same. No minds were changed. The *Miner*, the organ of W. A. Clark, voicing the opinion of the smeltermen, condemned "the apparatus as failing to accomplish anything." Officials of the Parrot told Hutchinson his "demonstration was not demonstration at all—that it was a failure."[10]

The city's enthusiasm for the messiah's "smoke consumers" had been tempered, but the conviction that a process could be developed to eliminate the smoke had been strengthened. Ruefully, the *Standard* commented that officials of the smelter "took especial pains to point out the imperfections of the scheme. . . . The result was that a good many went away disgusted and disappointed to find so many to explain defects and so few to explain the system itself."[11]

Most observers could believe only what they had seen: less dense smoke coming out of the stack than what they had been accustomed to seeing and breathing. Those members of the city council who were quoted reflected the consensus. "It exceeds my expectations," exclaimed the mayor, "and I believe the consumers should be put in every furnace of every smelter in town." "Today is the dawn of Butte's history," concluded J. W. Murphy, who promptly marked up his real estate 100 percent. Alderman James W. Forbis was satisfied that it got rid of the sight of the smoke; he could stand the fumes. And J. M. Stewart considered it "a great improvement." There were some, however, who thought Old Hutch more alchemist than chemist. It was "all imagination," declared Alderman Charles Goodale, and "I do not consider the scheme a success."[12]

Whatever the views of the city council, Health Officer Robarts took action. That evening he announced that he would proceed with the smelters as he would any other nuisance requiring abatement. He would notify them of the ordinance, and if they did not abate the nuisance he would arrest the superintendents and owners each day and fine them two hundred dollars daily until the nuisance was abated. "In all probability," the *Standard* commented, "the smeltermen will soon conclude that it is better to buy consumers."[13]

The council concluded that the "smoke consumer" afforded some improvement over existing conditions and that with some effort on the part of the smeltermen the means could be found to perfect it. Ordinance 167 was adopted to effect this end. The ordinance explained the council's viewpoint:

> After due inquiry through correspondence and the practical demonstration made by Mr. Hutchinson we have come to the conclusion that the smoke and fumes emerging from smokestacks and smelters constantly menacing the lives of people can be abated. We now, therefore, recommend that the Board of Health be

> instructed to enforce ordinance No. 167. . . . That it is a public nuisance . . . to erect, continue or use any building or other place for the exercise to any trade, employment or manufacture which by occasioning noxious exhalations, offensive smells, or otherwise is offensive or dangerous to the health of individuals or of the public.[14]

In addition, the council recognized a connection between the smelters dumping their waste on the town and certain citizens disposing of their garbage in public places. In another section, the ordinance declared it a public nuisance, subject to the same fine, "to cause or suffer the carcass of any animal, or of any offal, filth or noisome substance" to be deposited or allowed to remain "in any place to the prejudice of others."[15] With a flourish of the pen, the city council had codified a revolutionary measure. Henceforth, the community's health would be a common responsibility, enforceable by law.

The reaction of the *Standard* to the treatment accorded Old Hutch and his demonstrations by the smeltermen was of no small importance in moving the city council to enact the sweeping legislation. Even so, the heralded messiah, deterred from turning smoke into gold, left town dismissed as a false prophet and fuming at the smeltermen: "In all my experience I never received such malignant opposition, such a marked predetermination to make me fail."[16]

In a series of editorials after the tests, the *Standard* turned from disappointment to indignation. It censured the smeltermen for their failure to approach the experiment in "the spirit of courtesy and of patience which," it claimed, was "the usual accompaniment" for men seeking "in an experimental way after the truth." Accusing them of bad faith, the paper charged, "The courtesies of the Parrot smelter were offered to Hutchinson in order that he might be promptly disposed of. . . . [A]s soon as the average layman puts in a word for Hutchinson he is attacked with all the symbols found inside the two covers of a chemistry book, and that settles it." The *Standard's* "chief regret" was that "no satisfactory conclusions" had followed the experiment and the smeltermen had done "nothing to enlighten the people as to a remedy." Under the circumstances, it concluded, "The duty of the authorities is plain. Turn on the ordinance. Either the smoke or the people of Butte must go—and to say that the people will have to go if the smoke does is a venerable fallacy that frightens no one. Now is the time to act."[17]

The city did act. It sent abatement notices to all the smelters, informing them of the new ordinance and its requirements.[18] The Parrot smelter also acted. It closed down. Promptly, the *Standard* took to the ramparts on behalf of the cause. In a long editorial titled "Smoke War," it told the city the reason for the closure: "Undoubtedly it was distorting things to ascribe the closing of the Parrot smelter to the agitation of the smoke question, as some people assumed to do. The suggestion does not give reasonable credit to the business sense of the gentlemen who manage that property." The real cause for the shutdown was "perfectly well known to every businessman in the city." It was because of the prevailing stringency in the world's money markets: "The great copper companies [in Butte] are carrying stock aggregating millions in value for which none of them find a market. That is the condition of all of them from the Anaconda Company down to the most modest producer. . . . They are carrying a load which is burdensome for the best of them." Regardless of how "unfriendly the sentiment of the smelter people may be at this time," it was wrong to assume they would "inaugurate a policy of sulking by closing down their works." The "serious misfortune" of men "thrown out of work" and the "unwelcome interruption of business" was "unavoidable" because of the world money market, the editorial concluded.[19]

Then shifting to what it considered to be the heart of the matter, the *Standard* attempted to set the reigning philosophy of social Darwinism on its ear and provide the community with a larger voice in determining policy of the smelter operators. Even were a general shutdown to occur, the paper said,

> it would have little effect. Butte has made up its mind on this smoke question and has already adopted a policy. It is simply a question of the survival of the fittest, and public opinion has concluded, that, valued as they are in the aggregate of Butte's industries, the smelters that border the city . . . and the concerns that persist in heap roasting are not the fittest, especially in view of the growing conviction that they can continue in their prosperous career and at the same time adopt methods which will rid the city of a crying nuisance.[20]

The city council, however, was less confident than the combative newspaper. It reassessed its principle of common responsibility for public health and granted the smelters a reprieve to allow them to install abatement apparatuses.[21]

Some thought it a mistake to grant an extension. The *Standard* was deeply suspicious of the proposition that the smelters intended to try to eliminate the smoke. It felt the only agency capable of solving the smoke problem was the law. The town, it concluded, had financed one experiment, which the city council had unanimously considered at best a success, or at worst, an improvement. Yet, the smeltermen had judged it a failure and worthless. In view of this, the city's position against the smelters should harden; the city should not throw away "any more favors in this line."[22]

Throughout the month of February the paper focused on this theme. That February 1891 the city was relatively free from smoke, but pneumonia and influenza reached epidemic proportions. Saint Ann's and Saint James's hospitals were filled beyond capacity; Saint Ann's had to place mattresses in the hallways. Employers reported extensive absenteeism; physicians said that nearly all their patients were sick. Names were being added to the mortuary list at an increasing rate.[23] The *Standard* wondered about the source for such "an alarming amount" of sickness. Some citizens, it claimed, suspected the source could be found in the synergistic effects of the smelters' emissions: "While the smoke may not be visible, the timid ones contend that its gases and component parts are nevertheless afloat in the air, breeding sickness and disease, and demand that something be done to remedy the evil."[24]

The paper's editorializing reflected the fear within the antismoke movement that the campaign had faltered, that some of the reformers had lost their faith. Reminding the city council that it had given the smelters a limited time to curb their stack emissions, the paper pointed out that, "so far they have not made a move." The issue had "resolved into a condition of unconscious desuetude." No steps had been taken to rid Butte of the smoke since Hutchinson's departure. "It is evident," the *Standard* warned the council, "that the smoke cannot be bluffed out of Butte. Bluffs don't go in smoke wars."[25]

If Hutchinson's demonstrations were a failure, it continued, "it is about time some other experiments were being tried." "The average community," the *Standard* reported, "is usually afflicted with a short memory and the danger is that, as Hutchinson is gone and practically forgotten, so the source of torment will pass out of the public mind as early spring brings clear, bright days to Butte. . . . It is imperative, in view of the contingencies sure to arise in the immediate future, that decisive action be taken forthwith."[26]

The editorials could not have been more prophetic nor the anxieties better founded. Several days later, the clothes on Mrs. James Buckley's clothesline "turned green and were completely ruined." Captain John Lyons, a neighbor of the Buckleys, was forced to abandon his home during the night, and Shay, another neighbor, who had been fighting pneumonia, "took a turn for the worse." Then, early in March, the Butte and Boston and the Boston and Montana syndicates showed their contempt for the

Butte City Council and its ordinances by firing their ore heaps and smothering the communities of Meaderville and Dublin Gulch to the north in a stinking, gray blanket of smoke and fumes. Eventually fumes concealed the whole town. "Old Hutch's name was on every one's lips," a *Standard* reporter wrote, and "the curses hurled at the 'stink piles' and furnaces were awful in their fierceness and vehemence." After this "fumigation," the reporter pointedly suggested, the city council "may be spurred into making another effort to accomplish its abatement."[27]

Within days the townspeople were feeling the consequences. "Gruesome and unsightly funeral notices," impersonal symbols of personal losses, fluttered "from every available spot on Butte's principal streets." As the death rate continued to rise, the *Standard* voiced the thoughts of anxious citizens when it headlined a story, "Death Swings His Scythe and Cuts Down the Young and the Old." Even more despairing was the news that Butte's experience was "not shared by populous cities only a few miles away." And New York, readers were told, with a population over 4 million, had a death rate of 2.5 per thousand, while Butte's "unpleasant average" was 3.3 per thousand. Within a week after the heaps were lighted, half the police force was bedridden and "suffering severely."[28]

Once again the community read of people dying "strange deaths" like that of Fred Garing, who had collapsed at home months before. Accounts accompanying the reports of death were sparse, but most victims seemed to share three experiences: (1) They had died late at night or in the early morning hours; (2) they had complained "for some time" about "feeling poorly"; and (3) their sickness had been diagnosed as a cold or "consumption."[29]

During the first three months of 1891, between the time Hutchinson vowed to destroy the smoke and the time the heaps were "suffered to die out," 246 deaths were recorded at the city health department. In the nine months that followed, when inversions were infrequent, the total deaths for each month ranged between 29 and 46. During just two months—January and March, the months when respiratory and chest diseases were most prevalent and the smoke heaviest and most recurrent—40 percent of all "recorded" deaths for 1891 were logged. These months provided empirical proof of the validity of Dr. Robarts' assertions about the deleterious effects of the smoke—and its incapacity to destroy germs—to all but the most obtuse and those most dedicated to deceiving others. The statistics were crude, yet instructive. The city health department reported that out of the total of 246 deaths from various causes, 174 to 179, or about 71 percent, resulted from pulmonary-related diseases, such as lung abscess, chronic bronchitis, asthma, and croup. From pneumonia alone, 162, or 67 percent, died. Of those dying from pulmonary diseases, there was more mortality among people of long residence than among newer residents. Almost twice as many men died as women; miners composed 25 percent of the male dead. The average age of all deaths was thirty-eight.[30]

Shortly after the heaps were lighted, fifty residents of Meaderville petitioned the county commissioners to stop the open-air roasting. "The burning ore in the open heaps in the open air are [sic] a menace to the comfort, happiness and life [of residents] who heretofore having lived in a healthy locality, will be afflicted as bad as Butte has been for the last three months," the petitioners claimed. They identified Commissioner Charles H. Palmer, the superintendent of the Butte and Boston, as one of the gentlemen responsible for the heap roasting. In 1890, when the abatement movement was just getting underway, Palmer had told the smoke committee that "heap roasting knocked everything out as a producer of rank smoke." He told them then that he had stopped heap roasting and had also prohibited the Boston and Montana from doing it because the smoke was "so bad that half his employees were sick all the time." The men had become so "desperate" that they and some neighbors were "chipping in $1.00 to $3.00 to retain [lawyer] Charles O'Donnell and have him issue an injunction." Fortunately, Palmer had said then, "he had discovered their scheme just in time to foil it and prevent the inauguration of a mischievous retaliation of injunctions on the part of their neighbors."[31]

The commissioners, evincing the arrogance of potentates, rejected the petition. It had not been properly addressed, they said. It had omitted the name of Commissioner Palmer. Consequently, "it was an insult to one of the board" and altogether "too flip." The petitioners "must tone down the appeal." Commissioner John H. McQueeney told a reporter: "We want them to understand that they are not to insult one of our own number if they expect anything from us." Rebuffed but not subdued, "the boys on the hill" prepared another petition. Endeavoring to buoy them up, the *Standard* expressed what, for the time, amounted to heresy: "The people who are directly affected by the fumes from these piles, like the people of Butte, do not believe that there is any legal way any more than there is any moral right for corporations to build smoke piles under their noses, threatening their health, and rendering existence a burden."[32]

In the meantime the city had inched painfully to the *Standard's* position that the only way to stop the smoke was to arrest the smeltermen, a measure "acknowledged to be an extreme one." The *Standard* set forth the council's dilemma: "It is either legal or illegal for the smelters to endanger public health by filling the city with noxious fumes, and the present is a good time to ascertain whether the offense is legal or not. If the movement fails thru lack of law on the subject, it will then be time for the people to see to it that laws sufficiently strong on this subject are enacted."[33]

Whether it was legal or illegal, moral or immoral, there was no doubt among the smeltermen what they should do. They had no intention of "falling in line." They engaged in what Robert G. McCloskey in his history of the U.S. Supreme Court termed "the bad old American propensity to traduce the law whenever it conflicts with immediate interests."[34] Captain Thomas Couch, superintendent of the Boston and Montana, in the face of the expanding death rate, justified the practice in terms of rights and practicality: "Our people are willing to do anything to get rid of these fumes, but we would like to have any scheme practically demonstrated before we are compelled to expend a large sum of money on it. It is not right for the city to compel us to do so." If the arrests were carried out, the smeltermen warned, "every smelter in the camp will close down."[35]

The *Standard's* rebuttal was almost instantaneous. Speaking for the leaders of the movement, it replied, "If this is done, in the present state of health in Butte, such action on the part of the smelters would be a blessing. It would throw some men for a few days or weeks out of employment perhaps, but it would undoubtedly save human lives." The tactic, it explained, was to prevent further stalling: "The aim of those who are at the bottom of the movement to arrest the smeltermen is that they may be compelled to find their own remedy."[36]

The city prepared warrants for the arrests, but such extreme measures were averted at the last moment. The smeltermen told city officials that "it would be a good idea to turn the smoke question over to some expert of acknowledged ability and let him try his hand at it." Skeptics said no. It was just more procrastination. Discussion of the "various schemes" would proceed from "the present time until doomsday," and "Butte's afflicted people will simply be met months hence with the bland assurance that science failed to meet the emergency." And, if this were timed right, the skeptics said, it would be too late to undertake construction because of winter weather.[37]

The *Standard* had decided there was only one technical solution—the one "which no man has yet been able to dispute": high stacks. High stacks like the kind the Anaconda Company had constructed in Anaconda. They were the only thing that could "assure the speedy and unconditional removal of the scourge."[38]

The city agreed to delay the arrests, if the sheltermen would demonstrate "a disposition to try some remedy." Arrangements were made for a new "messiah" to be in Butte by March 23 to conduct experiments. In the meantime, the city suffered through the smoke, and Walkerville representatives met with Superintendent Palmer and in a "gentlemanly manner" reached an "amicable settlement." Palmer and Superintendent Couch of the Boston and Montana agreed to stop heap roasting and "suffer the piles to die out" at the end of the month.[39]

The new messiah was Col. Jacob I. Storer of New York, a sixty-six-year-old chemist and metallurgist. The colonel had achieved something of a reputation as an inventor. He held a number of patents for furnaces and garbage disposal devices. His scheme for "eradicating Butte's smoke" was patented under the title "A Process for the Treatment of Furnace Gases." It was designed to remove the sulphur from the smoke while at the same time saving considerable copper and silver, which the smelters were losing through the stack. Storer's process would convert the sulphur fumes into elemental sulphur, which could then be stored and perhaps marketed as fertilizer.[40]

Both Thomas Couch of the Boston and Montana and William A. Clark "spoke highly" of the process. According to one of the colonel's business partners, "Mr. Clark himself" believed they had the proper method and Couch was pleased with its "simplicity and efficacy." In an interview with the *Standard*, Colonel Storer did not look upon the forthcoming demonstration as an experiment. He had tried it in his laboratory and "it was a demonstrated fact." The problem would be in adapting it to the smelters' furnaces. The health department made arrangements for the use of the smelter at Clark's Butte Reduction Works, and once more the hopes of the town soared. Even the *Standard* took heart and remarked that "the friends of the smoke movement feel very cheerful over the present outlook."[41]

Dr. Robarts, in an address to the city council shortly before Colonel Storer's planned demonstration, manifested the satisfaction of the friends of the movement. They assumed the colonel's "model" would "prove acceptable to smelter authorities." Robarts praised the aldermen for their steadfastness in the effort to abate the smoke. "You have taken the righteous and just course," he told them, and he outlined how difficult the crusade had been and what a remarkable achievement it promised: "A clear air, free from impurities, that breathing may be easy, is the superior to all stimulants." The councilmen had made a commitment to something beyond the present. "Butte is not a temporary city. We have a generation coming for whom we are responsible," he said. It was their duty to "save the products of this city from foreign investment" and "from those who bask in other climes" and "remove the oppressiveness that wealth creates and the wealthy do not breathe, and we shall have the home and the fireside of which the generation we have made will bless us."[42]

Storer's proposed method had freed the city from the alarming and fearsome responsibility of implementing the law and bringing about a showdown with the smeltermen. They thought the battle had been won. Overlooking the *Standard's* injunction about bluffs and smoke wars, Dr. Robarts concluded: "To succeed in this without coercion and to preserve the interests and harmony of our people proves the value of time and subservience." "This," he said with swelling pride, "will be the first instance where a great public wrong is remedied without recourse to law."[43]

– Chapter Four –

"The War of Wealth against Health"

In early April 1891, after a week in Butte's environment, Col. Jacob I. Storer, the "new messiah," was felled by what was diagnosed as "catarrh of the stomach." After two weeks in bed, he was near death and had to be removed from the region. The *Standard* reported that "the apparatus at the Butte Reduction Works" was "in readiness for the test," but the demonstration was never made. Voicing disappointment, the *Standard* expressed the city's epitaph for the "New York Messiah": "What promised to be such a glorious success must, perforce, result in a failure, much to the regret of everybody in Butte."[1]

The new messiah's failure left the city with the unpleasant prospect of enforcing the ordinance calling for control of emissions at the source. In the meantime, spring came to Summit Valley, and its bracing breezes not only alleviated the smoke but sapped the spirit for reform. Having passed a law and threatened enforcement, the city in its innocence chose to trust to the conscience and benevolence of the smeltermen. The leaders chose to rely on the opinion expressed by the *Standard* shortly before Storer's abortive experiment: "It is to the direct interest of the owners of the smelters to ascertain what can be done." Ultimately, the outcome was certain: "There are plenty of men who believe they carry under their hats the final solution of Butte's smoke problem." If they were allowed to "fool with the question long enough, some man may in the end turn up with a plan that can be made to work—there wouldn't be anything strange in that." The law of the marketplace guaranteed success, the paper implied hopefully, because "if the practicability of any smoke-destroying scheme is once demonstrated in its application to conditions in Butte, there is a useful field for the invention and money for the inventor."[2]

If the townspeople chose to ignore the problem and to hope for the best, the town's intrepid iconoclast did not. After watching the activity of the smelters for several weeks after Colonel Storer's departure, Rev. John Rounder treated his flock to a sermon on smoke. Headlined in the *Standard* as an "elegant and common sense discourse," the text Reverend Rounder preached asked, "What has become of the smoke war? Where are the men who were in the van a few months ago? Is the case to go by default? Does anybody believe that the smoke problem will solve itself?" Endeavoring to prod city officials into action, Rounder predicted that "when next November comes the smoke will settle down on Butte as thick, as unpleasant, and as deadly as ever. The problem is not solved, and judging from the apathy that now prevails, all the agitation and excitement and expense of last winter accomplished nothing."[3]

Rounder went on to point out that in the two months since the Storer experiment had been aborted, the smeltermen had done nothing. City officials had played into their hands:

> They gained the victory in the staving off decisive action on the part of the authorities. . . . If the smoke problem is to be solved, the people of Butte must get a hump on themselves. The agitation must be kept up the year round or it cannot be solved. . . . Is the city willing to pass another winter in misery? If this is the case the present apathy of the people of Butte is criminal, for it means death to some poor residents of Butte next winter.[4]

The problem was not merely springtime apathy and a preference for averting hard decisions. It was something much deeper. The city's dilemma was seen in the tortured editorials the *Inter Mountain* had printed over the course of a week in November 1890 when the antismoke campaign was just getting underway. In its first editorial on the subject, the *Inter Mountain* had affirmed that the "humblest citizen" was "entitled at least to fresh air" and was "justified" in objecting to the arsenic and sulphur emissions the smelters could afford to abate. But several days later, the paper strained to assuage any investor's fears that the smoke campaign was a financial threat: "The man who asserts that the movement results from any hostility to the smelting companies is simply a pitiful mischiefmaker who ought to be locked up on the charge of being either drunk or crazy. If the abatement of the smoke is going to affect invested capital in Butte to its injury, there is no sane man in this community who would insist on any further investigation of the subject."[5]

It was an innocuous matter that could probably be resolved "with entire friendliness," especially "since the companies have always shown a pride in the growth and welfare of Butte. . . . We see no objection," the editor explained reassuringly, "to an interchange of views or to the making of some scientific experiments for the purpose of making this city a pleasanter place to live in. It is a subject that affects every interest of the community, and if the evil can be remedied without harm or loss to invested mining capital, and without any ill-will, there is no reason why a thorough examination of the matter should not be made." Just two days later, however, the *Inter Mountain* angrily attacked foreign investment and certain of the town's unconscionables: "The doctors and undertakers don't object to the smoke but the mass of the people have a very positive view on the subject and have a right to protect their lives whether Boston capitalists like it or not."[6]

The *Inter Mountain* had thus articulated the city's dilemma. The citizenry was caught in a simultaneous attraction to the antismoke campaign and a resistance to it. Unexpectedly and with great flexibility, they could embrace or repulse either position. Their ambivalence was due to an inescapable reality. Out-of-state capital was necessary for industrial and commercial development. The same capital would provide jobs and generate revenue needed to provide essential community services such as fresh water, streets, lights, transportation, sewers, and health and police protection for an expanding population. Above these services, and ultimately reliant on the same capital, lay the cultural activities cultivated citizens like Dr. Robarts considered essential to a sound community. It was the capital of those "basking in other climes" that the city had assiduously sought for at least a decade and that it intended to continue seeking. Fearful of alienating this capital, local leaders in varying degrees were wont to protect the interests of the distant investor and mine owner at the expense of the overall well-being of the community.[7]

Spokesmen for the mining industry never tired of reinforcing this disposition by continually reminding the community just where the mother lode resided if the city was to build El Dorado. The prestigious *Engineering and Mining Journal* gave its assessment of the place of smoke in Butte's future—and the reasons behind the agitation to abate it—in an editorial of January 2, 1892:

> Regarding the future of Butte as a mining camp, everything points to continued and increased prosperity for a great many years to come. As regards Butte as a city its prosperity must depend considerably upon its ability to get along amicably with the great companies operating here.
>
> At the present writing the city is suffering from the smoke from the smelters, and the real estate men, who think that if there were no smoke in Butte real estate would advance 50 per cent in value, are at the head of a movement to fight the Boston and Montana Company and to interfere with its smelting operations.

> It is to be deplored that these citizens carry sufficient weight to engender a feeling of hostility against the smelters among a section of the citizens of Butte.[8]

Besides the city being inextricably bound to outside capital, there were other forces operating on a more subtle plane to work against aggressive governmental action. The state was not yet two years old, and the extent and nature of the powers granted in the city's charter were dubious. To the extent city officials did recognize their power, they were reluctant to exercise it. The municipal government of that period was the embodiment of the laissez-faire doctrine: That government which governs least governs best. Butte grew without plan and with a minimum of control. Consequently, through default, direction fell to the most powerful forces within the community—mainly the dictates of industrial enterprise and private greed. But, most importantly, few in the mining city wished to initiate practices that might inhibit their own opportunity to be tomorrow's mogul. Too many ambitious men pictured themselves as a future Marcus Daly or William Andrews Clark.

While Reverend Rounder was trying to energize the city to make provision for healthy living, Capt. Thomas Couch, superintendent of the Boston and Montana, was making provision for the destruction of life. The role of Captain Couch in the "smoke war" calls to mind the comment Cornelius Vanderbilt had made some years earlier: "What do I care about the law? Hain't I got the power?" the shipping and rail magnate had said.[9] The "Captain," as Couch liked to be called, refused to see any relationship between smoke and funerals. He argued that the people who lived in the "lowlands" around Butte, "hidden from view by smoke," were healthier than those who lived "on the hill where the sun shines all the time in winter." He wanted the city council to "instruct" the health officer to make more studies through the coming winter in order that a "truthful statement of the death rate per cent of population may be ascertained." Couch argued that a "close investigation" would reveal what William Andrews Clark had vainly contended years before: "Burning sulphur" was a "partial disinfectant outside among the filth found in our valleys" and "not so injurious to Butte after all."[10]

Captain Couch's ostensible concern was the community's health; his actual concern was the amount of ore piled in heaps on the Boston and Montana's property. In spite of his claim that "our people are willing to do anything to get rid of these fumes," in spite of the promise to city officials to end heap roasting, in spite of Ordinance 167, Captain Couch and other smeltermen had gone on constructing heaps without any plans for stacks. This was later defended by the Boston and Montana attorney on the grounds that the ordinance prohibited heap roasting, not the building of heaps.[11]

Captain Couch's company, the Boston and Montana, was in the process of constructing a large and modern reduction plant at Black Eagle Falls, a few miles below the city of Great Falls, Montana. The plant was expected to be working by February 1892. Ore from Butte was to be shipped by the Great Northern Railway to the plant. Couch had apparently concluded there was much to be gained by concentrating as much ore as possible in Butte before having it transported to Great Falls, regardless of the suffering the townspeople might undergo.[12]

In late November 1891 the annual inversion set in and the dreaded smoke season prophesied by the *Standard* and by Reverend Rounder began. During the following week there were some periods when the gray cloud floated east and hung over the community of Meaderville, just beyond the city limits. In the first days of December rumors spread through the city that some smeltermen planned to fire their heaps before Christmas. The *Standard* assured the citizenry that the new mayor, Henry Mueller, had not authorized "any such proceedings" and they were without the approval of the city council. "When the ordinance against heap roasting was adopted," the *Standard* said, "ample time was given the smelting companies in which to prepare for new processes. . . . The authorities will be expected to see to it that the ordinance is respected to the letter."[13]

Reflecting the view of the city council, Alderman James H. Lynch announced: "I don't believe in letting the whole city suffer for the wishes of two or three men, no matter who they are. I will stand firm on the question of not allowing the resumption of heap roasting."[14]

On December 5, the city was informed that heap roasts were "ready for the fagot" and that "a concerted movement was on foot among the smelter people to revive heap roasting." Mayor Mueller dispatched the city marshal to investigate the matter. At the Honorable W. A. Clark's Butte Reduction Works, he found one heap readied for lighting and another under construction. At Captain Couch's plant, the marshal found six heaps ready for the torch. City officials who went to Clark's plant were told the heaps would be roasted before the first of the year. Smelter officials explained their reason: "These heaps were started before the order was issued that heap roasting should be stopped, and the building of no new heaps has been begun."[15]

City officials were stunned. The *Standard* expressed bewilderment: "Inasmuch as the order was issued by the council in December a year ago that no heaps should be burned . . . it is difficult to see how the piles could have been commenced." Furthermore, "the piles are only a few feet from the kiln, and it would not seem to be a very expensive matter to transfer the ore in the heaps to the kilns."[16]

Captain Couch told the mayor that it was "absolutely necessary for the Boston and Montana to resume heap roasting," and he could wait no longer than December 15. Mueller warned Couch he would be arrested as the representative of his company if the heaps were lit. "They have thrown down this gauntlet and we will have to try the case in the courts," Mueller said. "I suppose we might as well begin as soon as possible. I am disappointed in the matter. . . . I thought from what W. A. Clark told me that he would not allow any more heap roasting." As a last resort to avert a confrontation, Mueller wrote letters to Couch and Clark, almost pleading for a reconsideration of their decision:

> Gentlemen: I am creditably informed that you and your compan[ies] intend to reengage in roasting ores in heaps within the three-mile limit of the city of Butte. As you are informed and well know, the roasting in heaps within the limits above mentioned is strictly prohibited by an ordinance of said city, and you are hereby notified and informed and faithfully enforced by me as the chief executive officer of the said city, and I am assured that in my efforts to enforce the same I will have the hearty cooperation of each and every city officer and of the citizens of Butte generally.
>
> Hoping that I will not be compelled, by your actions, to enforce the provisions of said ordinance against your compan[ies] or any of your employes [sic], I have the honor of signing myself, Your obedient servant,
>
> H. Mueller
> Mayor of Butte[17]

The *Standard* interviewed the city aldermen. All except two wanted strict enforcement of the smoke ordinance. Of the exceptions, Alderman John E. Dawson thought it incredible: "I don't believe heap roasting is to be resumed. I saw the report in the papers, but I don't believe it. I think it must be a mistake." Alderman Eugene O. Dugan personified the ambivalence that had immobilized the city fathers since passage of the ordinance:

> The only objection I have to enforcing the ordinance is that it may throw men out of employment. . . . Butte has just had one shut down and it is not best to have another right away. I have no consideration for the smeltermen, but I don't like to see laboring men thrown out of work. I think Mr. Couch and the rest should abide by the ordinance. They have had notice enough, and as the council has given them the notice, they should be made to abide by it. It seems to me that the matter could be talked over and an

> agreement reached without the necessity of throwing men out of work. I don't believe in extreme measures, but I certainly would like to see the city well rid of the smoke.[18]

In the meantime, in daily editorials the *Standard* sought to embolden the mayor and galvanize the citizens to support him. "The people of Butte discover that they must make square issue with some of the smelter owners," the paper said. The smeltermen "were evidently of the opinion that the popular sentiment back of the [smoke] ordinance was merely a passing craze which would die out, leaving the law forgotten and its provisions a dead letter." Now some of the smeltermen "propose to rebel. . . . If there is going to be a contest," the paper concluded, "we ardently hope that it will be a fight to a finish. Life, health, comfort and the comparative value of millions of property, outside of smelting plants, are involved." The smelter owners had managed, one editorial claimed, "to silence merchants who enjoyed their trade. . . . They even found defenders in men of prominence in professional circles, they cajoled the owners of small properties with the cry 'no smoke, no wages for workingmen,' they were able to create and kept alive a strong public sentiment in advocacy of the idea that Butte cannot survive without enduring the smoke plague."

What did the city ask of the smelters? the editorial continued: That they "adapt themselves to improved processes for roasting. These are known to exist. Money must be spent, but the smelting business in Butte can afford to spend it." The editor asked that the city back up the mayor "from court to court, if need be, until Butte either wins decisively or is hopelessly whipped."[19]

Captain Couch was unmoved by the mayor's plea and unintimidated by the *Standard's* editorials. Shortly after a meeting with the mayor, with the insouciance of an executioner, the captain instructed his men to fire the heaps and then left town on the evening train.[20]

The fond ideals articulated, sincerely but ingenuously, by

Dr. Robarts—of the "real wealth of the people," of the value of subservience, of maintaining harmony by avoiding coercion, of remedying a public wrong without recourse of law—were incinerated on the ore heaps of Captain Couch. This time no messiahs promised mechanical relief. The dreaded confrontation had arrived.

Couch's action, however, outraged the city and momentarily blotted out fears of economic reprisal; it joined people of every social level in a cause the *Standard* labeled "the War of Wealth against Health." The usually cautious *Inter Mountain* headlined the most extreme comment yet made: "Heap roasting is nothing more nor less than attempted murder. That is the plain English of it."[21]

The day following Couch's departure, the city lay under a "stifling cloud of smelter smoke"; one could "scarcely see half a block." That evening the city attorney appeared before the city council to advise them on the law: "I have no advice to give except to enforce the ordinance. If we can't enforce our laws we had better find it out." The next day the city filed for an injunction against the Boston and Montana to stop it from heap roasting. The city argued it had the responsibility to protect the health of "all citizens and all persons" within its corporate limits and had the "power to abate and define nuisances" within those limits. The court issued a temporary restraining order and ordered the company to show cause on December 16—six days away—why a permanent injunction should not be granted.[22]

City officials, however, could not find any official of the Boston and Montana upon whom they could serve the restraining order. The smeltermen had "turned tail," cried the *Standard*, "and in the meantime the smoke rolls on and the injunction don't injunct." There was another problem. The city attorney pointed out that, even if the restraining order was served, it was doubtful anything could be done because no one could get close enough to put the heaps out.[23]

Meanwhile, talk of vigilante action spread throughout the town. Some citizens talked of forming an "expedition" and seizing the equipment of the fire department, raiding the Boston and Montana smelter, and flooding the piles. "It might be illegal but it would be damned justifiable," exclaimed one proponent. Mayor Mueller, however, who owned the Centennial Brewery, insisted on conforming to the law. Responding to a demand by some citizens that he lead a body of men down to the plant to extinguish the heaps, he contended, "I must act within the law. It would not do for me to head a number of vigilantes. I will do everything I can within the law, and if the citizens want to go to more extreme measures, they must act without me."[24]

About fifty citizens called for a meeting to be held in the city council chambers to consider means and methods for extinguishing the heaps. Attending the meeting were some of the most prominent citizens in town. F. E. Sargeant, a member of the board of trustees of the Anaconda Copper and Silver Mining Company, was named chairman of the meeting by acclamation. A number of respected residents addressed the assembly. Sargeant thought the law had been "insulted and defiled" and the people of Butte "outraged." Couch's action was "cowardly and contemptible," Sargeant said. "We have at least the right to breathe. The law can settle part of this matter; but I think our forefathers who threw the tea overboard in Boston harbor had not so much cause for action as we have here. They could get along without tea; but we must have air. People are sick and action must be prompt."[25]

E. H. Irvine, mine owner and one of the biggest real estate operators in the city, told the crowd the city was "justified in abating this nuisance if it has to abate Captain Couch at the same time." The community should take the law into "its own hands and protect itself," he said. "Every man's life is in danger by the act of Captain Couch. The community whose men will not protect their families and firesides is not worthy of the name of the community. . . . If our mayor and aldermen have not backbone enough to enforce . . . [the smoke ordinance] let the citizens take it in hand."[26]

Others expressed outrage at the actions of Captain Couch, but few were in favor of "mob action." Irvine asked the city attorney if there was "anything to prevent any peace officer from abating this nuisance." "The courts must first decide that it is a nuisance," the attorney said. Irvine replied that it seemed to him "great corporations were surrounded with a technique of the law, employing delays. . . . The people will stand by the officers in this action. It only requires a little sand in the city government."[27]

The next day a newly appointed "citizens' smoke committee" met with the mayor at city hall. W. McC. White spoke for the committee, telling the mayor they thought the city should enforce the ordinance and then let the courts decide on its constitutionality: "The smoke ordinance is quite as constitutional as many ordinances which are being enforced today and which have not been adjudicated by the courts," White contended. Mayor Mueller, however, maintained his position and insisted the city had to follow the "steps of the court." The heaps could not be "put out legally until the courts decide they are a nuisance."[28]

In lieu of a decision to extinguish the heaps, the mayor and the citizens' smoke committee decided to make their appeal to the very seat of authority. Mayor Mueller, on behalf of the city, and three members of the committee, on behalf of the people, telegraphed A. S. Bigelow, president of the Boston and Montana Company, in far-off Boston, apprising him of "the gross outrage perpetrated upon" their city "in violation of law and in cold and deliberate disregard of the health and lives of our citizens." The mayor wanted Bigelow to have "the opportunity of correcting the great wrong already done by at once ordering the discontinuance of this death-dealing nuisance before still greater injury and loss results."[29]

The committee informed President Bigelow that over two hundred deaths had occurred the previous winter, deaths that were "chargeable to the deadly pall that filled every niche and space in the city." Asserting that "self-preservation" was the "first law of nature" and that his company "in utter disregard of the laws" was "guilty of the greatest and most atrocious crimes ever committed in a civilized country," the mayor told Bigelow he must stop "at all hazards . . . this plague of death."[30]

The mayor's resolve to stay within what he conceived to be the law had weakened. He now seized the position that further action with regard to the heaps would depend on "a satisfactory answer" from Bigelow. At the same time, Mueller, as an individual, filed suit against the Boston and Montana, alleging that heap roasting was "injurious to health, offensive to the senses, and constitutes an obstruction to the free use of property and interferes with the comfortable enjoyment of life and property." He demanded a thousand dollars in damages and an injunction restraining the company from heap roasting. An indemnifying bond of ten thousand dollars, financed by ten of the city's most respectable businessmen, was posted. The court, once again, issued a temporary restraining order forbidding heap roasting until the "further order" of the court.[31]

The mayor could savor some satisfaction in the court's decree, but his dilemma remained unchanged: Who was going to enforce the decree? Alderman Anthony H. Barrett, feisty owner of an extensive harness and buggy manufacturing concern, supplied him with an answer, while at the same time revealing the reasons that sustained the mayor and other citizens with whom he held counsel: Barrett wanted Captain Couch and other officials of the Boston and Montana arrested, the heaps destroyed, and the company fined a total of sixty thousand dollars. "I shall do my best to have this done regardless of the trade I may lose," Barrett said. "Businessmen told me today that they did not dare do anything because they would lose trade. I need trade as much as any one, but I value my life and the lives of my family and friends more than a few dollars."[32]

While the mayor waited and hoped for a satisfactory answer from Boston, agitation and resentment grew until the city verged on a popular riot. Pamphlets turned up all over town with the inscription, "Smoke, Death and Destruction: Do you insist that the whole population of the city be exterminated?" Readers were asked to attend a meeting in front of city hall at noon the next day.[33]

In the meantime, people were leaving the city by the score. Those who could stand the expense were sending sick family members to other communities. All the rooms in Walkerville, situated atop Butte Hill and free of smoke, were sold out. "The rich are going to California or the East," reported the *Standard.* The Union Pacific and Northern Pacific ticket offices were experiencing record sales. For one day's business the two railroads reported ticket sales worth four thousand dollars. One of the most conspicuous citizens standing in line to buy a ticket to San Francisco for himself and his wife was Judge George W. Stapleton, "astute attorney for the Boston and Montana Company."[34]

On the appointed day, the Boston and Montana's attorney filed a petition to show cause why a permanent injunction should not be granted. The nucleus of the company's defense was total denial: It denied that the city had the authority to pass the smoke ordinance; it denied that heap roasting or any other smelting operation rendered "the atmosphere of said city unfit for or dangerous to be breathed"; and it denied that the "lives or health of the inhabitants" was endangered. In a demurrer the company even posed a philanthropic argument for heap roasting: "To shut . . . [the fires] off would throw men out of work," the petition claimed. Consequently, the Boston and Montana demanded dismissal of the city's complaint.[35]

Addressing itself to heap roasting, the court finally ruled. "When there is an infringement on the rights of others, that occupation becomes unlawful," the court said. "It is the duty of the court to enjoin it. . . . The temporary injunction is made permanent."[36]

Almost simultaneously Mayor Mueller made an announcement in an attempt to alleviate the mounting tension in the city:

> It is utterly impossible for me to withstand the entreaties of men whose wives and children are sick and steadily growing worse. Every 10 minutes in the day, some one comes to me with the story of others being sick. I act for my own family as well as for the families of others. It seems to me that self preservation is necessary and that the city will probably be depopulated in a few days if this horrible smoke keeps up. If I don't get an answer, and a satisfactory one, by 10 o'clock tomorrow morning [from Mr. Bigelow], we will move on the heap roasts.[37]

The mayor never did receive an answer from Boston. Instead, President Bigelow sent a coded dispatch to the chief clerk of the company. Late that night the clerk read what he "deemed polite" to the mayor and other dignitaries assembled at the Silver Bow Club, the most prestigious social club in town. The clerk could tell them only that Captain Couch would return to the city in about a week. In the meantime, the chief clerk was to "see the Mayor and get him to pacify the people."[38]

By nine the next morning a restless crowd of some two hundred men had gathered in front of city hall. Earlier, the mayor had attempted to disarm their anger by announcing that he had made arrangements with a contractor to begin covering the heaps. The crowd demanded "deeds not words" and remained "quiet but determined" until noon, when word circulated that ten wagons and 142 men were on their way to Meaderville to bury the heaps. Sometime later the sheriff's department was forced to post guards at the company's smelter to protect it from "knots of angry men" gathered there. A reporter commented on one sheriff's officer's wry remark to the mayor: "While no one a few days ago apparently had any responsibility for the acts of the company, now there was no difficulty in finding them. 'For God's sake cover up those heaps as quick as possible. Don't let that mob get out there!'"[39]

After working around the clock for three days, the men extinguished the heaps. On the day work commenced, the *Standard* had published records of the health office. It prefaced the statistics with the observation that "the smoke has been as thick as ever today; the death rate is appalling, and the undertakers are doing a thriving business." That day there had been five funerals. During the first nine days of the month, when the inversion had first trapped the smelter's stack emissions over the city, there had been eleven deaths. Since the heaps had been ignited, twenty-one deaths had been reported; for the preceding twenty-four hours, eight deaths had been recorded at city hall. Forty deaths were recorded within the city; how many had occurred in the county was unknown. On the day the heaps were extinguished, the *Standard* expressed relief. "Dense fog filled the streets, and if the sulphurous smoke had been added to the fog people would have choked. . . . It was a god send the heaps at Meaderville were covered."[40]

Captain Couch reached Butte two days after the heaps had been extinguished. He was met at the station by a crowd of "sympathetic and curious friends" who kept him "busy shaking hands" and discussing "his position and opinion" on the smoke question for which, the *Standard* reporter commented, "many people were inclined to hold him responsible."[41]

Suddenly, inexplicably, it must have seemed to its readers, the tone of the *Standard* had changed. Gone was the contemptuous editorial comment, the righteous anger, the condemnation, the insulting threats. In its new style the paper reported that Captain Couch "thought he had been accused unjustly throughout the whole business." He explained the reason for violating the city ordinance: "I was honest in my belief that the smoke from our heaps never came to Butte. . . . I had the new heaps built and fired to show that I was in the right, and it demonstrated I was mistaken."

If he had known how bad conditions had been, he explained, he would have "immediately ordered the heaps covered." He had "known absolutely nothing about what was going on here" and had not "run away from any anticipated trouble." He had been "ordered" to Arizona to inspect some mines and could not delay the trip. Was he going to burn any more heaps? the reporter inquired. "Certainly not," Couch said. "The people can rest assured of that. The company, nor I would not intentionally violate any law, and, as I said before, the heaps in this case were fired only to test the question whether the city or I was right in regard to the smoke."[42]

According to the reporter, one man standing nearby was unimpressed: "He knows that he was violating the law when he gave instructions to start those heaps, and all the explanations he can make will not explain that fact away," the fellow said.[43]

It had cost the city twenty-five hundred dollars to extinguish the heaps. Although Captain Couch affirmed that he "had always tried to do the right thing by the town and its citizens," the city attorney pointed out that "the company had committed an illegal act and the court's order was valid." Still, Couch did not believe his culpability extended to reimbursing the city for the expense of terminating the nuisance. Within days the matter was settled by an arrangement between city officials and the Boston and Montana. The complaints of the city and Henry Mueller were quashed in exchange for the smelter's confessing judgment to the original injunction and paying court costs. This would, according to city officials, make the injunction perpetual and save further litigation costs.[44]

Prior to the arrangement, a *Standard* editorial proclaimed, "Peace on honorable terms is in sight; pure air has made its conquest." Having declared the war over, the paper assumed the role of elder statesman and called for a magnanimous truce:

> There may be people in Butte whose plan it is to carp over the past and to pursue Captain Couch with their complaints. The *Standard*

> is not the organ of these people—when a man admits that he has been mistaken and offers to do the right thing, this newspaper is not disposed to crowd him. . . . We thought that the course pursued by the Meaderville management was dead wrong—we think so still; but what's the good in wrangling over it as long as Captain Couch himself admits as much? Rather let the assurance that Butte's plague is banished be the occasion for general rejoicing and the wiping out of old scores. Equip the smelters with proper appliances for roasting, build stacks—build them on a plan so liberal that the city is defended—and Butte will push ahead in the year 1892 at a pace that will astound the country.[45]

The *Standard* was either engaging in comforting self-deception, had honestly mistaken a battle for a war, or had recognized, however reluctantly, an implacable reality. Its abrupt change in its treatment of Captain Couch, irrespective of its rhetoric, suggests the last. Although calling bravely for a "liberal" building of stacks with abatement apparatuses, it carefully qualified the proposal by asserting that coercion would not be a part of the plan: "Plenty of time will be given in which to make the change," the paper said.[46]

The *Standard* had recognized, and the community had sensed, the reality of economic power. An individual community, regardless of how deeply felt its convictions, could not hope to contend with large corporations organized on a national scale and disposed to join together to exercise raw power to achieve objectives. Captain Couch was a symbol of that raw power and the response of the *Standard,* as well as of the constituency it represented, upon his return represents a capitulation to that power. The people could do little else. The necessary political, economic, and social structures were absent. Inexorably, they were forced to the state of mind that produced the *Inter Mountain's* assertion that only a "pitiful mischiefmaker" or a "drunk or crazy" man would jeopardize investment capital.

At bottom, the termination of heap roasting was a ceremonial victory. New techniques for concentrating ore had rendered it inefficient.[47] Bringing heap roasting to an end afforded but moderate relief from the smoke and fumes because of the ever-increasing capacity of the smelting plants. By the end of the next year, 1892, Butte's production in copper alone would increase as much as 60 percent over what it had been two years earlier. Over 40 percent of that increase was smelted in the valley below Butte. By 1900 production would increase again by 70 percent.[48] The same visitor, James Fergus, who had written his son in 1889 about not being able to see the town because of the smoke wrote him again in 1901. According to Fergus, there was little change in conditions in that span of twelve years, and he discouraged his son from coming to Butte because "the . . . smoke appears to settle bad this winter and there is a great deal of sickness."[49]

In the face of the hostile environment, the city's residents turned inward and bore their conditions with as good a grace as possible. They had come to accept the smelter emissions as inevitable. They could do little else. There was no way to force the smelter owners to allow those men who thought they "carried under their hats" the final solution to Butte's smoke problem to "fool with the question long enough." There was no longer a market for smoke messiahs.

Relief did not come until 1906 when the ore of the Butte and Boston and the Boston and Montana mines went directly to their smelters in Great Falls over a hundred miles to the north. Only the smelter of W. A. Clark and the Pittsburgh-Montana (Pittsmont) remained in Butte after 1906. By that time the Anaconda Copper Mining Company had bought out most of the other mining companies in Butte and was shipping the ore to their Washoe smelter situated some twenty-six miles away in the Deer Lodge valley.[50]

– Chapter Five –

Progress and Pollution Come to the Valley

UNTIL 1883 THE DEER LODGE VALLEY had been an idyllic mountain valley tucked almost a mile high between two ranges of the Rocky Mountains in western Montana. Extending in a northerly direction, it measured approximately fifty miles long and ten miles wide. Except for a few sawmills and some placer claims, only the measured work and muted sounds of a flourishing pioneer agriculture modified the valley's primitive nature.

Then, early in 1883, Marcus Daly, twenty-six miles away in Butte, began looking for a new smelter site to process the ores from his fabulously rich Anaconda copper mine. With characteristic foresight, Daly anticipated the water shortage in Butte that would eventually affect smelter production. On Warm Springs Creek, situated at the southwest corner of Deer Lodge valley, he found sufficient water for his prospective smelter. On June 25 he filed a plat for his city. Almost immediately the cows grazing indolently in the meadow retreated before the hundreds of workmen who arrived to build Anaconda, Montana Territory. Seemingly overnight an industrial city appeared in the valley. Within a month the population numbered fifteen hundred, and the city contained brickyards,

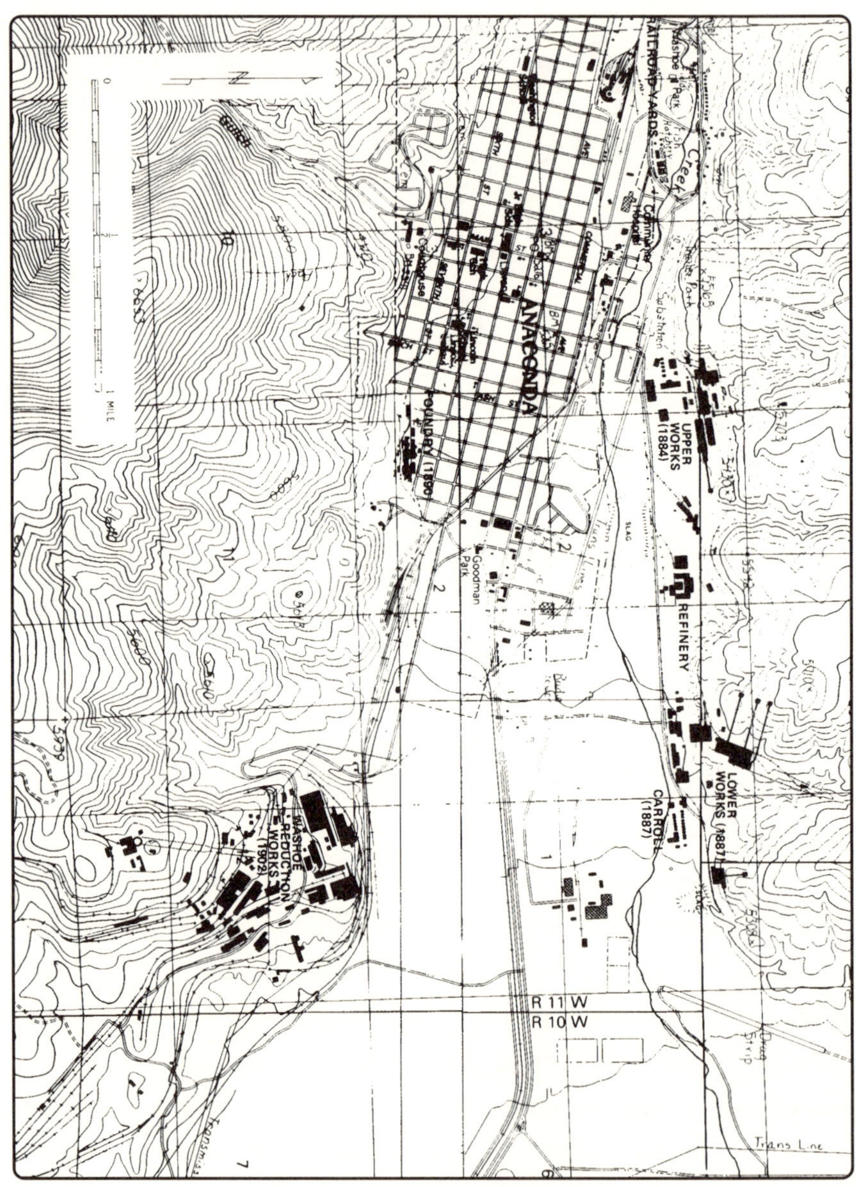

Industrialized locations in Anaconda.

BUTTE AND ANACONDA REVISITED: AN OVERVIEW OF EARLY-DAY MINING AND SMELTING IN MONTANA (N.P., 1991)

lumberyards, sawmills, and a variety of other businesses and shops that make up any thriving community.[1]

Daly built his smelter works fairly well into the mouth of Warm Springs valley on the north side of the creek. It eventually came to be called the Upper Works and could handle five hundred tons of ore a day. Within two years, improvements were made and the capacity doubled. In 1887 a new plant, having the capacity of three thousand tons a day, was opened about a mile away. Within the following fifteen years the smelting works underwent four successive plant expansions and technical improvements that added substantially to its productive capacity. Yet during this time no fume damage was done to the farming interests in the Deer Lodge valley. Apparently, with the smelters situated as they were, deep within the Warm Springs valley, the prevailing winds carried most of the emissions to the south and southwest where they were absorbed in the mountains and forest. The farmers did not become alarmed until 1902 when the Washoe smelter was built to refine the ores of the Amalgamated Copper Company.[2]

In 1899 the Amalgamated, with a view to centralizing control of the burgeoning copper industry, had bought the Anaconda Copper Mining Company, by then the biggest copper producer in the world. Anaconda's capitalization was at least $30 million. In addition, a host of other valuable mining properties in the Butte region was included in the purchase. Behind the purchase was the Standard Oil crowd: William G. Rockefeller, secretary-treasurer of Standard Oil; Henry H. Rogers, the canny securities manipulator and real power in the trust; and James Stillman, president of the National City Bank of New York, which was intimately connected with Standard Oil. Rockefeller and Rogers occupied seats on the board of trustees of the Amalgamated, and Stillman was a member of the executive committee. Marcus Daly was named president. Amalgamated functioned as a holding company, embracing all of the separate interests accumulated in the deal. Accurate data is difficult to come by, but some sources estimate the total capital of the combine at $400 million.[3]

The Washoe smelter was constructed to handle all the ore from the trust's Butte mines. Plans were made on a grand scale. It became the largest and most modern in the world, costing $9.5 million. It expected to cut the cost of refining copper as much as 2 1/2 cents a pound. Two-thirds of all the ore produced by the mines in Butte and vicinity, an estimated seven thousand tons of ore a day, were refined at the Washoe. To company officials, engineers, trade journalists, and local newspapers the plant was an engineering marvel. They boasted of the forty-eight "6-hearth MacDougal-Evans-Klepetke roasting furnaces," the five blast-furnaces, the fourteen "matting-reverberatories," the eight converter stands, and the four two hundred-foot-tall steel stacks that were lined throughout for discharging the smoke.[4]

Yet, with all the imaginative and sophisticated technology packed into the plant, only minimal thought was given to the nature of the product that would come out of the stacks and what its final disposition would be.

A few months after the furnaces were first fired in 1902, the farmers in the Deer Lodge valley began noticing peculiar symptoms of disease in their stock. Moved to distant pastures, those animals with milder cases effected a rapid recovery, but upon their return to home pastures, they experienced a recurrence of symptoms. The fact that most of the diseased cattle were found in those sections where the smoke drifted most frequently told the farmers that the Washoe smelter was responsible for the disease. "The Washoe Smelter belches forth such an enormous volume of these poisonous gases that the mind is incapable of comprehending the vast polution [sic] of the atmosphere in its vicinity," declared the leader of a protesting group of farmers in the Deer Lodge valley.[5]

The Washoe was built on the south side of Warm Springs valley on a ridge extending down from the foothills of the surrounding mountains and projecting into the juncture of the Warm Springs and the Deer Lodge valleys. At that point the prevailing winds carried more smoke into the Deer Lodge valley than in former years. As much as fifty-nine thousand pounds of arsenic trioxide and notable quantities of sulphur dioxide, copper, antimony, lead, zinc, and other substances were discharged into the air daily.[6]

Operations had begun at the Washoe on January 25, 1902. By November, carcasses of several hundred horses and cows lay scattered over the various ranches in the valley. At one ranch sixty carcasses, mostly horses, lay in the corner of one field. At the Callan Brothers ranch nearly the entire herd was cut down. A year later, a drayman in Anaconda fed his horses for a week on hay grown during the 1902 season. One went into convulsions and died. A postmortem examination showed that various tissues of the animal and its liver contained 13 parts per million of arsenic trioxide. An analysis of the hay fed to the horses revealed 285 parts per million of arsenic trioxide.[7]

The farmers secured the services of two chemists to conduct an investigation, Professors W. D. Harkins of Montana State University in Missoula, and R. E. Swain, of Stanford University, Palo Alto, California. Animal deaths reached a peak in the period between September and November 1902. The professors conducted an extensive investigation and determined the reason for the acceleration of fatalities. The fresh grass of the spring had little time to accumulate arsenic from the air, and the summer period was a time of rapid growth and frequent rains, but with autumn came a dry period during which there was no plant growth. Consequently, it was during autumn that solid substances adhered to the plants at a rapid rate. In order to protect their stock, the farmers were forced to move them out of the district and import hay for those that remained.[8]

The farmers also suffered losses at the marketplace and in land values. Peter Levengood, a farmer and rancher in the valley since 1867, testified that since the beginning of the Washoe smelter's operation his hay had decreased 75 percent in quantity and "pretty near the whole in quality." Because of the arsenic content, Levengood said, "If I sell a man a load of hay I can never sell him another." His land, three miles northwest of the smelter, previously worth fifty to seventy-five dollars per acre, was now worthless: "No one will make an offer . . . at all," he complained.[9] Nick Bielenberg, one of the biggest ranchers in the valley, with fifty-eight hundred acres situated twelve miles from the Washoe, testified that since the smelter had begun operations, he could not graze stock on his land but had to move them many miles through the mountains to the Big Hole basin. In the autumn of 1902 and early months of 1903 he lost about a thousand head of cattle, twenty head of horses, and about eight hundred sheep from the effects of the smelter fumes. Many of the survivors had become "sickened and unthrifty." The only time Bielenberg could utilize his lands for grazing was during the growing season, when rainfall washed the poisons from the grasses. Breeding on his range had to be terminated, and he no longer operated at a profit.[10]

The principal objective of the farmers in securing the services of Professors Harkins and Swain was to establish the cause and source of the disease. The professors tramped over one hundred square miles of country surrounding the smelter. They were in the area when the fatality among the farm stock was greatest. They dissected a "very large number of the more recent fatalities" and found that practically all of them evidenced arsenical poisoning, either acute or chronic. They analyzed vegetable tissue throughout the Anaconda area and found that within a quarter mile of the smelter tissues contained as much as 1,551 parts per million of arsenic trioxide and 1,800 parts per million of copper. As far away as four miles 88 parts per million of arsenic trioxide and 708 parts per million of copper were found in the tissues. In the tissues of cattle as far as three miles east of the smelter 11.3 parts per million of arsenic trioxide were found, in a horse five miles south southeast 2.8 were found, in a steer three miles northeast 8.6 parts per million, and in a horse three miles northwest 4.7. Some livers of animals contained from a trace to 63.12 parts of arsenic trioxide to 1 million parts of tissue. In a series of ongoing experiments the chemists were unable to establish definitely what amount of arsenic would kill a farm animal. It depended on a number of variables—the environmental conditions extant in the area as well as the condition and history of the animal. But it was apparent then and was later established that all soluble arsenic-containing compounds, as well as the element itself, are extremely poisonous and can cause death in farm animals.[11]

Two local veterinarians called upon to treat the diseased stock and inspect the dead agreed with the professors' findings of arsenical poisoning. The state veterinarian, Dr. M. E. Knowles, inspected the cattle and corroborated the professors' and veterinarians' observations, adding that only at Great Falls, the location of the Butte and Boston and the Boston and Montana smelters, and in the Deer Lodge valley had he ever observed the condition.[12]

Upon the complaints of the farmers to the smelter company, some threatening suit unless damages were paid, the Washoe selected two employees and the state veterinarian to investigate the claims and assess the damages. J. A. Dunlap was put in charge of finalizing negotiations and concluding settlement. Ultimately, the company paid $330,000 in damages for the year 1902. Many farmers, however, felt that the method and means employed in reaching a settlement were unfair. The company refused to pay damage for crops that were not within five miles of the smelter. The farmers complained that they had no voice in the selection of the three men making the investigation and assessing the damages; that to obtain any settlement they had to sign away all rights for future damage; and that they were told to "go to court" if they were not willing to take what the company offered. The farmers were reluctant to go to court, however, for fear of an unfriendly jury or one under "undue influences."[13]

Yet many of the farmers would not settle. One was Nick Bielenberg, whose experience and feelings may be deemed typical of those resisting. Bielenberg had submitted an itemized statement amounting to thirty-one thousand dollars for his damages. The company representative, Mr. Dunlap, responded by asking, "How would a check for ten thousand dollars look to you?" Bielenberg took the offer as an "insult." He requested that they arbitrate the damages "in any fair way," but the company refused arbitration and would not move beyond the ten-thousand-dollar offer.[14]

In 1903, between July and September, the Washoe smelter was closed for the installation of a new flue and stack system. Ostensibly as a result of the conflict with the farmers, the management turned to a solution that the *Standard*, over a decade earlier, had claimed was one theory "no man has yet been able to dispute."[15] Officials decided to deliver the dust, fume, and gases five hundred feet higher into the air. The fumes from the various departments of the plant were conducted through branch flues into one huge brick dust main that climbed twenty-three hundred feet up the hill behind the smelter and emptied into an enormous brick stack three hundred feet high and thirty feet in diameter. A tunnel was driven underneath the dust main where the settled flue dust could be dumped into ore cars.[16]

The function of the extensive flues was to allow the dust and some of the fumes to settle in the dust main before going up the stack. As reported in the company press and various trade journals, the principal reason for the building of the huge flues and stack was to prevent fume

Washoe Reduction Works, Anaconda, Montana, 1903.
MONTANA HISTORICAL SOCIETY PHOTO ARCHIVES, HELENA

damage to the countryside, or as one account phrased it, "to prevent the air within the 'smoke zone' from becoming impregnated with substances deleterious to the growth of vegetation." In addition, an arsenic plant was attached to the flue system; it removed two tons of arsenic daily, which was then marketed as fertilizer.[17] Theoretically, as a result of discharging the smoke from the new stack, eleven hundred feet above the valley floor—"the largest stack in the world"—natural diffusion would act upon the smoke stream in such a way that, by the time its poisons reached the ground, they would be so diluted as to be practically harmless. But whether the chief concern was settlement of the flue dust to avoid environmental damage or recovery of the flue dust to improve internal efficiency, whether the "remodeling" was primarily environmentally oriented or an economic one is debatable.

Indeed, evidence supports the economic motive. A. J. Mathewson, general superintendent of the Washoe plant, told a special assistant to the U.S. attorney general that prior to the remodeling of the plant he had the denuded mountainsides immediately behind the smelters "swept for flue dust in order to resmelt" the dust for its copper content. As a result, the special assistant emphasized to his supervisor, "he took out something over *six million dollars worth of copper*. This will serve to give you some idea of the enormous metallic waste from a smelter plant."[18]

Edgar M. Dunn, an engineer in the testing department of the Washoe reduction works, read a paper before a meeting of the American Institute of Mining Engineers in Butte in 1913 titled "Determination of Gases in Smelter Flues; and Notes on the Determination of Dust Losses at the Washoe Reduction Works, Anaconda, Montana." His long and technical discussion implied that the construction and maintenance of the flue system was almost exclusively a commercial proposition. In contrast to the lengthy title, the theme was astonishingly concise: "In a nutshell, how efficient is our flue system as a value catcher?" At the same meeting, an associate observed that "the Anaconda metallurgists recognized the fact that in copper smelting the copper values are nearly all contained in the dust, and designed their flue system to catch the maximum amount of dust."[19]

The cost of the new flue system was reported as $750,000.[20] Figures for the amount of metals—that is, copper, silver, and gold—recovered in the flue system of the Washoe and resmelted are not available, but a rough idea of the savings and ultimate profit on the investment can be found in the experience of the Amalgamated's Great Falls plant. Writing in 1915 about the savings that could be effected by an adequate flue system, a respected mining engineer, Charles H. Fulton of the U.S. Bureau of Mines, described what had been accomplished at the Great Falls smelter:

> It was demonstrated at the Great Falls plant that with the old flue system in use there was a daily loss in the dust of 3,775 pounds of copper [and] 106 ounces of silver. . . . [With] adequate provision for dust recovery, the losses have been reduced to 347 pounds of copper, 28.5 ounces of silver, and 0.2 ounces of gold. The copper loss amounts to 0.19 per cent of the copper charged into the furnaces. The saving thus affected [sic] is $372.18 a day, or $130,263 a year. The cost of the new flue system was $1,100,100 so that the saving made much more than pays the interest charges on the investment.[21]

At the Great Falls plant the dust and fume discharged daily was 50,690 pounds; the Anaconda plant discharged 153,554 pounds of dust and fume daily, or almost three times as much as did Great Falls.[22] Clearly, the new flue system at the Washoe was an investment upon which a return was expected and ultimately made, not a cost incurred in the interest of a clean environment in order to prevent damage to crops, forest, and animals.

In the course of the remodeling of the Washoe, many settlements were reached with the farmers, and, apparently as a kind of conciliatory gesture and an attempt to establish salubrious relations with the agricultural community, the Washoe sponsored a banquet for the farmers in the valley. Reportedly, William Scallon, president of the Anaconda Copper Mining Company, made a speech in which he promised ideal neighborly relations in the future: "We are satisfied that no more damage will accrue to you after we commence operation with the high stack; but if any should occur, come to us, show us your damage, and you will be compensated for the same."[23]

When the smelter reopened in September 1903 one could climb to a strategic point on Mount Evans behind the smelter and high above the valley and view the theory of "disseminating and scattering by air currents at a great height over a large area of country" the fumes, gases, and dust contained in the smelter smoke. In that normally clear and bracing atmosphere, the smoke stream pouring out of the stack a thousand feet above the valley could be seen trailing northward down the Deer Lodge valley for thirty miles toward the town of Garrison, or flowing eastward toward Butte, or swinging around to the south and sweeping up Mill Valley and filling the narrow ravines leading down from the Continental Divide fourteen miles away. The region surrounding the smelter is an area of variable winds that often blow with great velocity. Consequently, the area affected by the smoke was not confined to a narrow strip or circumscribed around the smelter. Each season has its prevailing winds, but they were not constant enough to contain the direction of the smoke to any considerable degree. Frequently, in spite of the elevation of the stack, the smoke could be seen settling into the valley less than a mile from the smelter, whereas on quiet days it might hang over the valley as a haze.[24]

In October 1903, the month following the renewal of operations, a sheepherder who lived and pastured his sheep in a small valley twenty-eight miles from the smelter and somewhat sheltered from the smoke by a range of intervening hills concluded it was time to move the band to better feeding grounds. After leasing a low field situated approximately fifteen miles northeast of the smelter, he drove 3,500 head into the lush lowland grass. A week later a number of sheep became lethargic, reclining on their sides and reluctant to get up. Then they became sick and refused to eat. Neighbors advised the sheepherder to drive the herd to a feeding place farther from the smelter. The fellow did, but on the way 500 sheep died. During the next four weeks 125 more died. Dr. M. E. Knowles, the state veterinarian, investigated the case and concluded that the sheep had died from acute arsenic poisoning. When a complaint was filed, smelter officials contended that the sheep had died from alkali poisoning. Professors Harkins and Swain were asked to investigate. An analysis of the soil and of the stomachs of the sheep determined that alkali was not the poison. A visit to the field revealed that grass cropped very close to the ground by the sheep contained 52 parts of arsenic per million, and some moss in the lower part of the field contained 405 parts per million. It was not clear, however, whether the arsenic was deposited in the field prior to 1902 or after the smelter opened.[25]

There were no serious complaints from farmers in respect to arsenic damage until late in the autumn of 1904. Then the symptoms of arsenic injury began to appear again. The animals that sickened or died were treated or autopsied by two Anaconda veterinarians. The state veterinarian was called in again, and all three identified the problem as smelter-fume injury of the same character the stock had suffered in 1902. The state stock inspector for the district agreed with the findings. In cattle the damage took the form of garlicky breath; ulcerated noses, with nostrils sometimes swollen shut; rough scruffy hair, with inability to shed; inflamed and weeping eyes; loss of appetite; constipation; diarrhea at advanced stages; loss of flesh; exhaustion upon exertion; tucked-up abdomen; abortion; and incapacity to breathe. In horses the symptoms were similar but with the addition of partial paralysis of the hind legs, labored heart action, dilation of pupils, and partial paralysis of the diaphragm.[26]

About the time these investigations were taking place, the Washoe officials pronounced the new flue system and high stack a "success" in controlling smelter fumes.[27]

Once again, the farmers, this time organized as the Farmers' and Dairymen's Association, began making calls upon Washoe officials. They sought "an amicable adjustment of the differences." They were met, they complained, with evasion and noncommittal pronouncements. On January 30, 1905, the association wrote a letter to the company, setting forth a "statement of facts." They employed two veterinarians, Dr. Charles Gresswell of San Francisco, who had experience in adjusting fume-injury cases in Swansea, Wales, and Dr. Olaf Schwarzkopf, a U.S. Army veterinarian stationed at Fort Assinniboine. And once again, the ranchers called Professors Harkins and Swain into service.[28]

Harkins and Swain began with an inspection of the new flue system and stack at the smelter. Following a series of complicated experiments to determine the velocity, volume, and weight of the gases, fumes, and dust moving through the flues, they concluded that

> While the great flue may be fairly efficient in settling the copper from the smoke, still a considerable amount escapes, while the amount of arsenic given off is very great. . . . The flue dust at the top of the big flue is roasted to produce a commercial arsenic trioxide, but all of the rest is sent to the reverberatories and smelted for copper, so that the arsenic which it contains must in the end be eliminated in the smoke, excepting only that portion which goes into the reverberatory slag.
>
> It is evident that a large percentage of the oxide of arsenic which condenses to the solid state while in the flue, is carried along simply because it is so fine that it will not deposit at the rate the gas is moving, *11.4 feet [3.47 meters] per second.* . . . The flue collects the greater part of the heavy metals, but allows most of the white arsenic to pass as a finely divided solid, and as a gas.[29]

On the effects of the high stack in lieu of the lower type, they found that

> The change has been to decrease the amount of arsenic deposited near the stack, and to spread it over a greater area. On account of the higher specific gravity of the smoke, it tends finally to fall to the earth, even though it may first be carried to a very great height, often striking the ground at distances of from one to four miles. As a general thing, the sulphur dioxide is more dilute when it reaches the ground than it was in the case of the lower stacks. However, there is often a greater amount of damage done in certain valleys which were previously protected by the mountains, but into which the smoke can now drop. The trees on certain high mountain sides

> are also acted upon more than before, in some cases showing a streak in the forest where the current of smoke passes, while below this the damage is less noticeable; but usually the gas finally descends into some valley. At times, in a high wind, the smoke will come to the ground at the base of the stack; and the leaves of shrubs in the valley about half a mile from the stack, and 1000 feet lower than its top, show a very high percentage of arsenic. . . . On level land, near the stack, the action of the sulphur dioxide is less intense than before, and as a result it is easier to raise trees and vegetables.[30]

In undertaking the study, the professors, while yielding nothing, apologized for trespassing in an area traditionally reserved for the engineer: "In sampling and analyzing smoke, among the most difficult problems is one which may seem to belong to the engineer, but which, our experience has shown, cannot be let to the ordinary man in this profession."[31] Judging from the results after 1903, they were not only correct about the role of the ordinary man in the engineering profession, but they were about a decade ahead of the mining engineers of the day. Writing in 1915 on the effect of the velocity of the smoke stream on dust discharge, Charles H. Fulton of the Bureau of Mines noted,

> Formerly it was accepted as a fact that the passing of the smoke stream through a long system of flues of such cross section that the velocity did not exceed 20 to 30 feet per second would cause the dust to settle. It was considered that length was essential. In the more modern construction it is found that if dust is to be settled the velocity must be reduced to not more than *2 1/2 feet per second* in clear chambers or to 5 feet per second in chambers hung with wires.

Fulton also agreed with the professors' conclusions on the effectiveness of high stacks.[32]

Although the Washoe officials had pronounced the new flue and stack system a success, the professors had shown that they were allowing the gases, fumes, and dust to course through the flues at a velocity of as much as four times the speed that would allow effective settling of the dust. They found that the average amount of arsenic trioxide thrown off in the smoke was 59,270 pounds per day—a little above that of 1902—and that 447,600 pounds of sulphur trioxide and 4,636,000 pounds of sulphur dioxide, plus a host of other metals and gases, were being discharged daily as of January 1905.[33]

Other findings showed that arsenic had settled in the livers and other tissues of animals in levels equal to and exceeding amounts found in animals in 1902. In October 1903 the liver of a horse pastured 1.5 miles northeast of the smelter had 13 parts per million of arsenic trioxide; the highest found in 1902, before the "remodeling," was 11.3 parts per million in a liver of a cow three miles east of the smelter. In the months from January to May 1905, substantial amounts of arsenic were found in animals situated in various directions from the smelter. In a cow's liver five miles northeast 10.3 parts per million were found; in another, 11.9; in the kidney of a mare fourteen miles north-northeast, 13.3 parts per million were present; in another 4.5 miles distant in the same direction 35 parts per million; in the lung of a gelding five miles directly east 7.1 parts per million. In the same period, arsenic was found in the grass and other vegetation at various distances and directions from the smelter ranging from 10 to 405 parts per million.[34] The farmers were as bad off as before. Their hay was left standing in the stacks because the dealers in Anaconda began shipping hay in from the Bitterroot valley at an average cost of four dollars per ton more than what the Deer Lodge farmers were asking.[35]

When the farmers went to the smelter to complain of damages, the company made investigations but determined there was nothing to the complaints. They paid no claims after July 1903, the date, ironically, of President Scallon's generous speech. One farmer who had suffered severely reported that A. J. Mathewson, general superintendent of the Washoe, told him that if he paid his claim he would have to pay all claims. Mathewson maintained that the farmers' claims were "groundless" and based on "what you might call cupidity." When the association suggested the claims be settled through arbitration, the company objected on "technical grounds." Apparently, the Amalgamated, faced with the prospects of paying damages indefinitely, of investing in nonprofit-oriented research to curb the poisonous emissions, or of buying the farmers out at sellers' prices, had decided to fight. The farmers were told to "go to the court house."[36]

– Chapter Six –

The Farmers versus the Trust

At the Willow Glen schoolhouse on the evening of February 21, 1905, 107 farmers of the Deer Lodge valley joined together to form the Deer Lodge Valley Farmers' Association in order to prosecute their claims against the Amalgamated. The association included owners of over 90 percent of the agricultural lands in the valley. These men felt that, having settled in the valley before the smelter was built, they had "prior rights" they thought the Amalgamated should respect. Not all farmers joined; some had claims or suits pending from the operation of the smelter in 1902, which they preferred to prosecute independently.

Acting through an executive committee, the association offered to settle all damage claims for about $1,175,000. This would include title to 60,525 acres, all the land owned by the farmers in the association, together with all water rights and improvements. Included in the settlement figure was the value of all the farmlands affected, which was estimated at $918,147. The farmers explained that the reason they had not made the claim earlier was because the company had promised, and they had believed, that the construction of a large stack would prevent further injury. Consequently, they had waited until they were fully satisfied as to the results of the new operation. Now, they argued, it was apparent that stock raising and farming could not exist in the valley so long as the smelter was operated on such an extensive scale, especially in view of the class and character of ores produced by the Butte mines. W. D. Harkins and R. E. Swain had concluded that, since the ore of the Butte mines had such unusually high concentrations of arsenic, and the concentration was increasing as the mines went deeper, under the present smelting methods at the Washoe the prospect for a reduction in the quantity of arsenic emitted was "not encouraging." The members of the association agreed to be off their lands by November 1, 1905, if settlement was made and the money paid by May 1, 1905. Otherwise they would be forced to sue.[1]

No reply was forthcoming. Thus, the farmers prepared for a fight with the trust. They raised forty thousand dollars within the organization to institute the suit. The agents of the trust scornfully warned them they would fight them in the courts until they "did not have enough money left to buy breakfast."[2]

Suit was filed in the federal district court in Helena on May 4, 1905, against the Washoe Copper Company and the Anaconda Copper Mining Company. W. H. Hunt was the presiding judge. Attorneys R. L. Clinton of Butte and C. M. Sawyer of Anaconda represented the farmers. Suit was filed in the name of Fred J. Bliss, the only nonresident property owner in the valley who could maintain a suit in the federal court. The two mining companies were Montana corporations and the farmers wanted the suit tried in federal court, where they felt more secure from "undue influence."

Through Bliss the Deer Lodge Valley Farmers' Association sought to establish their contentions in court. The suit was a "test case" but a lethal one, seeking to obtain an injunction permanently restraining the Washoe smelter from operating on the grounds that its emissions, particularly in the form of sulphur dioxide and arsenic, settled on crops and forage, damaging them and poisoning the livestock that fed upon them. In addition, the suit maintained, the same thing occurred on other farms situated in the Deer Lodge valley within an area designated as the "smoke zone." The smoke zone was defined as an area encompassing over one hundred square miles of improved farmland on which all the farmers suffered damage from the smelter. If the smelter was allowed to continue operation, the complaint argued, the valley's soil eventually would become so impregnated with poisonous substances that the land would be completely worthless for agricultural purposes.[3]

Fred J. Bliss lived in Emmet, Idaho, and was not a member of the association. He had lived in Butte until poor health forced him to sell his butcher shop and grocery store and move to a healthier climate. In lieu of cash, he had accepted 320 acres of farmland as partial payment on the business. The land was situated about five miles southwest of the Washoe. Title to the land was passed during the remodeling of the smelter in the summer of 1903. Bliss never lived on the land but rather rented it to adjoining farmers. As a result of the smelter's emissions, he claimed, the farm's rental value decreased from a thousand dollars per year to three hundred, and the land itself, originally worth twelve thousand dollars, had suffered sixty-four hundred dollars' worth of damages.[4]

In their complaint, the association's attorneys also listed other grievances. All the land within the smoke zone, claimed to be worth more than $2 million, had been rendered less profitable by the smelter fumes. Prior to the smelter emissions of 1902 the valley had been a rich and fertile farming country, well watered and well adapted to raising a variety of livestock and farm produce. The fumes had killed nearly all the trees in the smoke zone and had rendered the country adjacent to the smelter "barren and desert like." Ultimately, the farmers claimed, their homes would be destroyed and they would be compelled to migrate.[5]

The Washoe smelter was a prima facie nuisance, attorney Clinton argued, "where the pollution of the atmosphere with sulphur and arsenic is the greatest ever found in the history of copper smelting." The court should compel the smelter to eliminate poisonous substances from the smoke before releasing it into the atmosphere. "It is the duty," Clinton declared, "of defendants to devise ways and means to prevent pollution of the atmosphere."[6]

Cornelius F. Kelley, secretary of the Anaconda Copper Mining Company, an uncommonly capable attorney and sophisticated executive, directed the defense for the Washoe. Kelley answered the farmers' complaints, first by claiming that the Washoe had, "regardless of expense," remodeled the plant "in order to prevent the air within the 'smoke zone' from becoming impregnated with substances deleterious to the growth of vegetation." This was done, Kelley claimed, because farmers charged that the arsenic, sulphur, and other substances were damaging their crops and livestock. In addition, as a result of the allegations of damage, the company had paid $300,000 in claims. "Although they were not satisfied that any considerable damage" had been caused, "nevertheless [the Amalgamated had] compromised and settled the claims." The company had always been "willing, without litigation, to pay all damages" that had been "or may be sustained" from its operations, Kelley maintained, whenever "reasonably satisfied" that damage had occurred.[7]

Denying the plaintiff's claims about the Deer Lodge valley, Kelley argued that it had never been "rich or fertile" and did not contain one hundred square miles of improved farming land. For that matter, it did not even contain "any large area capable of being profitably used for growing wheat or garden products." The soil was for "the most part thin and poor and in many places charged with alkali." In some part, he admitted, it was "fairly adapted" to growth of farm products and livestock and some farmers were "thriving and prosperous." Nevertheless, the value of the land and property in the so-called smoke zone was not worth $2 million but at most $500,000. Most importantly, there were not "any considerable or injurious quantities" of arsenic or sulphur dioxide or any other noxious or poisonous substances deposited upon the Deer Lodge valley or on any land described as the "smoke zone." The land had not been "rendered less profitable or useful" by smelter emissions.[8]

Kelley warned of the grim economic consequences an injunction would mean to Anaconda, Butte, and, ultimately, the state, nation, and humanity. The smelter was responsible for the existence of the city of Anaconda, a community of ten thousand or twelve thousand people directly dependent upon the smelter for jobs and commercial opportunity. Its citizens had made investments of hundreds of thousands of dollars in homes and public buildings. Butte, a city of sixty thousand, or about 20 percent of the entire population of the state, depended directly or indirectly on the continued operation of the mines, and the Washoe was responsible for treating three-fourths of all the ore mined in Butte. Living within Butte and Anaconda were eight thousand men "absolutely dependent" each month for their livelihood on the operation of the smelter and its mines. More than $7 million a year was paid in wages and over $4 million in natural resources was consumed by the company's operation. Since the company's mines furnished Silver Bow County with 25 percent of the total taxes collected by the State of Montana, an injunction would "seriously impair the revenue of the state and render impracticable the continuance of the local county." If the company was enjoined, national and worldwide distress would follow because the smelter produced 20 percent of the nation's copper and 11 1/2 percent of the world's. The smelter's copper was used principally in the manufacture of electrical appliances and machinery and in the manufacture of various metals required by the country's burgeoning industry. A shutdown would mean a copper deficit in world markets that would bring on higher prices.[9]

A business has a right to pollute, Kelley argued: "Our position about that matter has been that there is no legal objection to a pollution of the atmosphere until it results in damage to somebody, which gives him the right to formulate a cause of action or to complain. We have a perfect right to carry on a legitimate business, and if incidentally we should pollute the atmosphere nobody has the right to complain until specific damage gives him a cause of action."[10]

Within weeks of Kelley's arguments suggesting the overwhelming importance of the Washoe Smelter—as compared with the valley's farms—to the economy of Montana and asking the court to balance the economic interests of the farmers against the preeminent interests of the smelter, a federal district judge in Salt Lake City, John A. Marshall, handed down a decision that must have struck alarm in the hearts of Amalgamated executives. In the case of James Godfrey et al., representing about four hundred farmers, versus the American Smelting and Refining Company, the United States Smelting Company, and the Utah Consolidated Company, the federal court issued an injunction designed to prevent any of the companies from treating ore containing more than 10 percent sulphur in the natural state or from permitting the escape of arsenic from the stacks. In his decision, Judge Marshall set forth his conclusions on the action and effects of sulphur dioxide and arsenic:

> This gas is heavier than air, and when cooled, falls to the ground at a distance from the smelters dependent upon the air currents. When it is brought in contact with moisture, either in the form of rain, freshly irrigated ground, or the moisture present in growing plants and the foliage of trees, sulphurous or sulphuric acid is formed, which is destructive to vegetation. Besides the emission of gas, some flue dust is emitted from the smelters which contains perceptible quantities of arsenic resulting in the death of horses and cows.[11]

The decision forced the smelter companies in Utah, at great expense, to smelt at a distant area where agriculture was not affected.

About the same time, a state judge in Solano County, California, rendered a decision in *Solano County v. Selby Smelting and Lead Company* that permanently enjoined the smelter from emitting fumes that could destroy vegetation and injure livestock in the county.[12] Attorney R. L. Clinton took note of the decision by Judge Marshall in the Salt Lake valley case and argued that the same defense of economic dominance was resorted to by the defendants. He quoted comparative figures and concluded that the comparative injury of closing the smelters in the Salt Lake valley would be practically as great as from the closing of the Washoe smelter. Therefore, the argument of economic preponderance of the mining industry made by attorney Kelley had little relevance to the law in the Bliss case. An area more suitable for smelting should be found for the Washoe, Clinton argued.[13]

The heart of the Bliss case focused around the testimony of "experts." In the Progressive Era, roughly the period between 1890 and World War I, there were basically two opinions about the role of experts in American government and society. Progressive thinkers, typified by Theodore Roosevelt, conceived of experts as agents of popular government employed to regulate objectionable business practices. The work of these experts would eventually lead to supplanting the then-current pernicious system of government by big business. Other progressives, represented by Woodrow Wilson, viewed the expert as a mere hireling of big business and vested interests. One could not expect effective regulation from those controlled by the very interests that were the object of regulation. The only way to beat such a combine, they thought, was to return government to "the people." The work of the "experts" in the Anaconda smoke case, as it soon came to be called, supplied ample support for both views.[14]

When the farmers initiated the suit, they had not appreciated just how soon it would be before they could "not afford to buy breakfast." They were, for instance, unable to secure the number of expert witnesses the Amalgamated produced. They admitted, the *Inter Mountain* told its readers, that they were unable to hire the services of any experts "except under very unsatisfactory assurances and at a very large compensation to be first paid them." Their counsels wrote a number of toxicologists, pathologists, and histologists. All but one refused to get into the case, "some pleading other business engagements" and others candidly confessing that they were "not willing to enter the field against the defendants."[15]

Money was obviously not the only problem. Something else was preventing experts from testifying on the association's behalf. The farmers realized that Amalgamated's influence reached into high places, and when one of their veterinarians, Dr. Charles Greswell, an expert from Swansea, Wales, died on the operating table in Denver before giving testimony, they suffered some disappointment. When, however, Dr. Olaf Schwartzkoph, their other veterinarian, who was stationed at Fort Assinniboine, was transferred to a remote part of the Philippine Islands before the trial, their suspicions were stimulated.[16]

Harkins and Swain, nonetheless, continued their investigations into 1907. They found "striking proof" that the arsenic was deposited by the smelter smoke and not absorbed from the ground as the Amalgamated attorneys claimed. In Mill Valley, about ten miles southwest of the smelter, grass samples taken in 1906 contained from 48 to 583 parts of arsenic trioxide per million. At the Bliss ranch, north of the smelter, samples taken at the same time contained 18 to 73 parts per million of arsenic trioxide. Hay cut from the Bliss ranch in August 1904 contained 30 parts to the million; by the next April the grass in the same field contained 263 parts, or almost nine times as much as it had eight months earlier. Numerous other examples were detailed. "This is deposited arsenic," the farmers asserted, "rather than absorbed arsenic" because "by shaking dry grass or hay grown in the vicinity of the smelter a finely divided dark-gray powder, running notably higher in arsenic than the tissue from which it came, is obtained." Thirty miles north of the smelter, at the town of Garrison, 35 parts of arsenic trioxide were found in the grass.[17]

Anticipating an argument, the two chemists wrote, "It may be claimed that the arsenic and copper found in these samples are not deposited from the smoke stream, but absorbed from the soil itself." In the Deer Lodge valley this was definitely not the case, they said. Plants do not absorb so much arsenic and copper from the soil. The two chemists had grown cereals on several soil samples collected from the Deer Lodge valley, taking them outside the range of the smelter smoke and planting in them barley and timothy. Substantially lower amounts of arsenic were found in these plants at maturity than were found in similar samples planted in the same soil five miles from the smelter. Structures similar to highway billboards were set up at various points around the smelter at distances of up to five miles. Vaselined cloths covered the fronts of the structures. Some faced the smelter, others faced away, and still others were situated facing upward. Substantial amounts of arsenic were found on those facing the smelter and significant amounts were found on the others. "The results admit only one interpretation," the professors declared, "which is that the smelter smoke is the source of the arsenic found in such excessive amounts in the vegetation of the region about Anaconda."[18]

To establish the morbidity and lethalness of the effects of arsenic on animals, experimental verification was made by feeding animals graded doses of arsenic and analyzing the tissues. Dissection of the animals showed the presence of arsenic in every organ. In one case the ulcerated nose of a horse showed the presence of arsenic to be as high as 1,015 parts per million. Four of the healthiest sheep were chosen from a flock and each fed different daily doses of arsenic along with arsenic-free hay.

Overall, in the first week the sheep gained weight, but by the end of seventy days all but one had died. It too would have died if the experiment had not been terminated. The experiment showed that 46 milligrams of arsenic trioxide per day to 100 pounds of body weight were sufficient to kill a sheep. Arsenic—a carcinogen to both humans and animals—was found in the milk of cows that fed upon the grasses as far away as six miles, ranging from 1.08 grains to the gallon to 34.00 grains to the gallon.[19]

At the direction of the Department of Agriculture, Dr. Robert J. Formad, chief pathologist and histologist of the U.S. Bureau of Animal Industry, and Dr. E. T. Davison, veterinarian of the U.S. Department of Agriculture, investigated the livestock of the valley and corroborated Harkins and Swain's findings. Dr. D. E. Salmon, founder and chief of the U.S. Bureau of Animal Industry for twelve years and one of the few experts the farmers *could* bring in, also confirmed the professors' conclusions. On the witness stand he asserted, "I think the opinions of the experts for the defendant companies to the effect that the animals in the Deer Lodge Valley were not suffering from the effects of arsenical poisoning, were absolutely incorrect."[20]

Amalgamated imported a battery of experts to controvert the findings of the farmers' experts and to "show that every precaution had been taken by the Washoe people to prevent injury to surrounding industries and to minimize the danger." The company's experts were reportedly paid as much as fifty to a hundred dollars a day and were comfortably ensconced in the Amalgamated's fashionable Montana Hotel in Anaconda. They were, said the farmers disdainfully, mere "college professors from the East."[21]

One of the defendant's witnesses was the superintendent of the Washoe himself, A. J. Mathewson. The press observed for the benefit of their readers that Mathewson, "having shown that he knew all about smelters," testified that the precautions taken at the Washoe to prevent damage were "greater than any place he knew of except possibly some small plant located right in some city." In his judgment, it was not possible to take any further precautions and none were necessary. The Washoe's method for controlling emissions was the "natural way," Mathewson testified: Dilute them with air from high stacks. An expert had recently been sent to other countries and had gathered "all the information there was to secure on the subject," he told the court. It was "the universal testimony of all that the high stack, disseminating the smoke through the air, is the only practical method, the very best that science has to offer."[22]

To buttress Mathewson's testimony, the company brought in two experts from New York—Charles A. Doremus and Frederick J. Falding—and Profs. F. W. Traphagen of the Colorado School of Mines and W. W. Strange, a meteorologist. Traphagen, Doremus, and Falding were "company partisans" who had testified in other fume-injury suits on the side of smelters. Doremus was a "consulting chemist" employed by the Amalgamated in New York. He had once made an "authority's report" on a smelter in Tennessee without ever having visited the site. He had a reputation as an "eminent toxicologist and laboratory chemist." In the summers of 1905 and 1906 he had made an investigation of the stock conditions in the Deer Lodge valley and the nature of the emissions of the Washoe for the Amalgamated. He concluded, after a "detailed study," that the exit gases were not "proving poisonous to livestock" and that "damage to livestock was negligible." Two years later, employing figures provided by the company, he drew the same conclusion without visiting the smelter or the Deer Lodge valley.[23]

Falding was a chemical engineer employed by the Amalgamated at an allied plant in Tennessee. Although he had never been to the Washoe plant, his testimony was defined as "expert" on its operations. His conclusions were based on figures supplied by the company. At the

Tennessee smelter he was in charge of the operation of converting fumes into sulphuric acid for the fertilizer market. He affirmed Mathewson's theory of the "natural way" and declared that "the methods in use at the Washoe smelters [for the control of emission] are the most complete and efficient of which he had any knowledge." The methods employed at the smelter in Ducktown, Tennessee, to extract sulphur dioxide from the emissions, he said, would not work at the Washoe because of the size of the western plant and quality of the sulphur dioxide gas.[24]

Professor Traphagen, with heavy interest in mining and smelting, testified as to the volume of sulphur dioxide in the Deer Lodge valley as a result of the emissions of the Washoe. He produced data from Golden, Colorado, to show that the amount of sulphur dioxide present in the air in Golden was more than it was in the Deer Lodge valley and the vegetation at Golden was not injured. Therefore, he attested, the vegetation in the "smoke zone" could not be injured from the amount of sulphur dioxide emitted by the Washoe. Counsel for the farmers investigated his data and on cross-examination forced him to admit that the aspirators he used at Golden to collect the samples were situated around an assay laboratory running thirty-two coal-burning furnaces, the coal of which released high concentrations of sulphur dioxide.[25]

The meteorologist, Professor Strange, let his enthusiasm carry him away in strengthening the theory of the "natural way." On the stand with notes in hand, he testified that between the hours of 11:30 P.M. and 1:05 A.M. the wind's velocity over the smokestack ranged from 81 to 134 miles per hour. He had, however, neglected to take the air pressure during the blow. Plaintiff's attorney Clinton pointed out to the court and the professor that the highest wind velocity ever recorded in the United States was 87 miles per hour. In the world, he told the embarrassed expert, the highest wind current ever recorded was 92 miles per hour—a typhoon in the Indian Ocean.[26]

The experts testifying for the farmers were at a decided disadvantage in the case. The farmers, having neither the wealth nor the national influence of the Amalgamated, had to rely on chemists Harkins and Swain and on the experts of the Department of Agriculture. The latter were primarily interested in forestry and soil erosion problems, and consequently most of their investigations had been confined to the national forests, that is, forest land owned by the people of the United States. Only in a peripheral way could they relate to the farmers' damage. In addition, the farmers' authorities testified independently and upon their own original and separate investigations, the findings of which were, in many cases, new and unorthodox.

Amalgamated's counsel utilized a different approach. His experts testified under the novel "joint preparation" approach. Under cross-examination they were almost impregnable to plaintiff's attorneys. They were all titled gentlemen, all swathed in respectability and authority. Professor Theobald Smith, bacteriologist of Harvard University; Prof. V. A. Moore, Cornell University; Prof. Leonard Pearson, University of Pennsylvania's state veterinarian; and Dr. Duncan McNab McEachran, Canadian Government Veterinarian Service, all assembled in Anaconda and Helena to overcome the testimony and evidence of Professors Harkins and Swain and Dr. Salmon on the effects of smelter fumes on farm animals. None of the defense experts had any previous experience with smelter-fume injury to livestock, but by virtue of their positions and reputations they seemed above reproach. They were, Circuit Court Judge Erskine Mayo Ross pointed out, "men of national, and some of them international, reputation in respect to the subjects upon which they testified, and the positions respectively held by them are good evidence of their personal character."[27]

The farmers, however, saw something different: "No one appears here for the company but wealthy or high salaried men," one of the

These images of healthy crops and livestock, circa 1906, are from the Anaconda Company's "Smoke Damage Series" and were presumably submitted as evidence in the Bliss case. The series also includes close-ups of healthy horse nostrils and photographs of intestines infected with parasites—evidence that arsenic was not the primary cause of damage to livestock in the Deer Lodge valley.

MONTANA HISTORICAL SOCIETY PHOTO ARCHIVES, HELENA

association members observed, "all talking for self-interest as well as for the companies [sic], and none of them advance any way to pay the damage, and only say: 'Don't close the smelter.'"[28]

The Anaconda Copper Company's veterinarian, Dr. Robert Gardiner, a young man recently graduated from veterinarian school, had been put in charge of conducting the overall investigation for the defense. Earlier he had concluded that the stock in the valley were suffering from lice and starvation. Dr. Knowles, the state veterinarian, said he had never seen a bad case of lice in the valley. Samuel Vaughn, an employee of the copper company, testified he had seen only one "lousy" cow in the valley and had sprayed and cured it.[29]

Under the direction of Dr. Gardiner, the experts "from the East" began their investigations. The methods they employed in these investigations and the remarkable consistency of their findings amazed the plaintiff's attorneys. Although distinguishing themselves as scientific men, their method was that of imitative scientism. Questioned by attorney Clinton about the sixty-six autopsies performed, Dr. Duncan McEachran disclosed the method of "joint preparation" employed by these expert witnesses:

> *Clinton*: Well, sixty-six autopsies, and there was no material deviations or disagreements among the four or five of you?
>
> *McEachran*: No, there was nothing—oh, we had occasionally a little talk, but nothing you could call a disagreement.
>
> *Clinton*: Then my understanding is that when Dr. Theobald Smith was present, in conducting a post-mortem, that he practically prepared the notes in those cases, and you and Dr. Gardiner assented to them?
>
> *McEachran*: Yes.
>
> *Clinton*: Was there any case in which you all took notes of the

post-mortem changes, whereupon comparison you found that all your notes agreed?

McEachran: No, there was no case where we all took notes that I know of. I might explain, Mr. Clinton, that we had so little time and so much to do, and were so often interfered with by weather, snowstorms, and cold, etc.

Clinton: But one party would take the notes and you would practically all agree to the notes?

McEachran: Yes, we all heard the description. If Dr. Moore was describing the stomach we all were looking at it and surrounding him, listening to it, seeing what he was describing, and the notes were being taken down in the presence of them all. If Dr. Pearson was describing a liver, the same thing occurred.

Clinton: Then his description would be incorporated and made part of the notes?

McEachran: Yes.

Clinton: Then, I state that during all of these numerous autopsies and post-mortems that you made, you don't recall of any material disagreement among any of the parties?

McEachran: No.[30]

Another aspect of the method of "joint preparation" in organizing the company's defense was described by the dean of the group, Dr. Theobald Smith. Attorney Clinton questioned Smith about how the group's testimony was synthesized:

Clinton: Then they all agreed to your notes and your dictation; there was no disagreement?

> *Smith*: I don't believe that question came up; I simply dictated the notes, because I made the study. If they accepted them they will probably answer that; I have not asked them.[31]

Dr. Gardiner was selected to, or the chore fell to him of, presenting the experts' findings in court. None of the others would swear to the entire presentation. Clinton observed for the court that it was strange that Gardiner should swear to the mass of typewritten testimony as his own when he had so little to do with preparing it. This, declared Clinton, made Gardiner "a mere phonograph witness."[32]

The manner in which Dr. Gardiner's testimony was prepared was yet another side of the "joint preparation" approach. On the witness stand, Dr. Gardiner answered the queries of Clinton as follows:

> *Clinton*: Where were those notes prepared that you are reading from now?
>
> *Gardiner*: These notes were prepared in Mr. Kelley's office.
>
> *Clinton*: By Mr. Kelley?
>
> *Gardiner*: By Mr. Kelley's stenographer; they were copied from a written paper which I handed to him, and they were corrected by me.
>
> *Clinton*: When were they prepared?
>
> *Gardiner*: About three or four days ago.
>
> *Clinton*: What have you there, are they carbon copies?
>
> *Gardiner*: These are the originals in almost every case. I believe there were a few copies which I have not got the original. In the majority of instances I had the original copy.
>
> *Clinton*: Were they not all original as far as the information contained?

Gardiner: Yes, sir, they were.

Clinton: What was the purpose of having more than one copy made?

Gardiner: These copies were joint notes and duplicate copies were made for the purpose of supplying each of the persons making the autopsy with a copy of the autopsy notes.

Clinton: Now, in making these notes at the time you say they were joint notes?

Gardiner: They were joint notes.

When each of the defendant's witnesses went upon the stand to testify, he carried with him a copy of these joint notes and from time to time referred to them before answering questions involving the statements of other experts in prior testimony. Attorney Clinton objected to the procedure: This was merely "a prepared case read into the record." It was "merely the concurrence of dummies" to the assessments of Harvard's Dr. Theobald Smith. Judge Hunt did not sustain the objection.[33]

One of the most pathetic representatives of academia to testify in the Bliss case was a puffed-up, self-important professor from the agricultural college in Bozeman. Professor Joseph Blankenship testified for the Amalgamated. Thereafter he became "a well-known professional witness," appearing frequently for various smelter companies being sued for damage resulting from fume emissions.[34] Professor Blankenship conceived of himself as a scientist ahead of his time. He had discovered the disease responsible for the damage to vegetation in the Deer Lodge valley—the "drying up disease," he termed it.

Clinton: Doctor, where did you ever find this drying up disease, until you came to the Deer Lodge Valley?

Blankenship: I never saw that disease before, if you refer to the one immediately around the stack.

Clinton: Can you refer me to any scientist, such as botanical or florist people, that have ever isolated or identified that disease?

Blankenship: To the best of my knowledge . . .

L. O. Evans: I object to that, it is not proper cross-examination. If counsel desired to examine the witness upon the recognized authorities that is one proposition, but to ask him if he can refer him to any scientist, that is hardly proper.

Clinton: Well, I will ask him that way if you prefer it. Doctor, I will ask you, do you know of any authorities; could you refer me to any recognized [authority] that deals or treats of this drying up disease that you discovered in the Deer Lodge Valley?

Blankenship: I may say that I discovered that disease. I am authority for that, and I think that future scientists will approve what I say. It is not a discovery that I have generally made known, but I believe it is the true reason of the dead growing around the stack there.

Clinton: It would be something like the Jones method?

Blankenship: Yes, it might be; I don't know what the Jones method is, though.

Clinton: Or the Traphagen method?

Blankenship: Every scientist in making investigations in any department of science, must, if he wanders beyond the domain of human knowledge, discover new things, and that is one thing I believe that I have discovered.

Clinton: You never saw it anywhere else?

Blankenship: I saw it at Great Falls [the location of Amalgamated's other smelter].[35]

The plaintiff's attorneys thought the most striking proof of the morbid effects of the arsenic was the ulcerated noses of horses. Harkins and Swain argued that "undoubtedly" the ulcers formed as a result of feeding on grasses with high concentrations of arsenic, causing the arsenic dust to lodge in the folds of the nostrils and irritating the mucous membrane. In some cases the nostrils were closed so completely that the animal had to be destroyed. Dr. Salmon produced the same type of ulcer by applying arsenic dust from the Washoe to the nostril of one of the afflicted horses. The defendant's experts argued that the arsenic could not cause the "sore nose" disease. Yet they were unable to produce a similar ulcer by any of the agents they claimed were responsible for it. This was not only true of the sore nose disease but of their entire defense, the farmers' attorneys argued: "Their whole defense is one of conjecture and doubt." The defense experts could "positively testify to nothing as a fact concerning the abnormal conditions of the livestock and crops."[36]

The "vital point" attorneys Clinton and Sawyer thought the case should turn on and the argument the defendants and their experts had not been able to overcome was that

> Whatever injuriously affects livestock in the smelter area affects at the same time and in substantially the same manner (making allowances only for different anatomical structure) all species of livestock subjected to this widely distributed cause of injury in the Deer Lodge Valley.
>
> No contagious or infectious or any other disease affects all species of animals in the same manner or at the same time. When you stable the animals and remove them from exposure to this widely distributed cause they all improve. Stabling and changing feed has

Caricature of L. O. Evans, joint counsel for the Amalgamated Copper Company in the Bliss case.
CARTOONS AND CARICATURES OF MEN IN MONTANA (N.P., 1907)

never been a remedy for any form of disease that would affect all species of animals.[37]

The "nauseating testimony" of the defense experts was "beyond the human mind to fathom," the plaintiff's attorneys claimed, expressing at once their disgust and the farmers' incredulity. The defense amounted to obfuscation, attributing all damage to animals and crops to everything but the smelter fumes. L. O. Evans, joint counsel for the Amalgamated, remarked at an unguarded moment that "We have not been able to contradict your analyses showing arsenic in the Valley."[38]

The case, although containing many knotty and diverse technical questions, in the end reduced itself to a consideration of the knowables, attorney Clinton pleaded. Take into consideration the "law of nuisance" and what was known about the operation of copper smelters where ore containing large concentrations of copper sulphide and arsenic was smelted; take into consideration the known action upon animal and vegetable life of arsenic and sulphur fumes as evidenced by the samples and testimony produced in the trial; consider that "it will not be necessary to accept a single new theory, but merely apply the common knowledge of all mankind to the facts in the case," Clinton said, and "we believe the members of his Honorable Court can safely, quickly and positively arrive at the facts in this case."[39]

Judge Hunt rendered his decision on May 26, 1909. His findings of fact were strikingly similar to those advanced by attorneys Kelley and Evans in their final argument. Most significant among them was that the operation of the smelter and its mines constituted the chief industry of the state upon which the cities of Butte and Anaconda were almost completely dependent. A better location for the smelter could not be found. The "only possible conclusion from the evidence" was that the Amalgamated had built its plant in accordance with the "best known methods and processes" for rendering the emissions harmless. With the remodeling of 1903, they had done all they could to make the plant "consistent with the state of the art."[40]

The judge deemed the testimony and experiments of Profs. W. D. Harkins and R. E. Swain as "inconclusive," "too indefinite and uncertain" to overcome the findings of the court. While recognizing that "without doubt great damage was done by the smoke" before 1903, since that date the smelter emissions only affected the Bliss property and that of other farmers in the smoke zone to a "very slight degree." No "substantial injury" had been done to the crops or lands in the valley since the building of the big stack, Judge Hunt concluded. There was, however, evidence that timber on the hills back of the smelter was being damaged. But apart from this "special damage in this immediate vicinity, the weight of evidence is to the effect that the principal damage to timber was done before the new stack." Considering the "purely equitable defenses" and the "magnitude of the interests involved"—the $10 million cost of the smelter, resource consumption and wages, taxes and state income, and responsibility for 20 percent of the U.S. copper supply—Judge Hunt denied the injunction. Bliss's "special damages" were assessed at $350. On appeal, the appellate court in San Francisco confirmed the decision, on the grounds that to do otherwise would operate contrary to the "real justice of the case."[41]

The decision stunned the farmers. Real justice had not been served, they felt. According to the chairman of the Association's executive committee, Walter Staton, the concept of "magnitude of interest" was "a bulwark behind the necessities of the public so as to escape the consequences of . . . unlawful acts." "This is as legitimate," Staton said angrily, "as the fellow who was being sentenced for an attempted murder, who said, when asked any reason why sentence should not be passed: 'Why yes, your honor, if I go to jail my family will suffer for sustenance,' and the Amalgamated would like to claim all Montana as its family." "But when they want to close," he added, "close they do," and he listed four instances when Amalgamated had shut down its works for periods of from several weeks to over a year within the space of six years for reasons unassociated with the internal operation of the smelter. "It is the dear people of the state when they want to smelt, and the people be d——d when they don't want to," Staton fumed.[42]

It was the *Inter Mountain*, however, that probably reflected the feelings of most Montanans. Editorializing on Judge Hunt's decision, the paper fairly sighed with relief, saying that "for more than three years the smoke case has acted as a drag to the prosperity of the Deer Lodge Valley, and except for the disappointed farmers" everyone could now rejoice that the smoke case had ended.[43]

– Chapter Seven –

The Struggle outside the Court

While the lawyers were waging battle inside the courtroom, the fight went on outside. Unlike the farmers, the Amalgamated waged its campaign on a number of fronts—the economic, the social, and the political—mainly through their press. In the court battle the trust had to assume the defensive, but in the battle outside chambers they went on the offensive, forcing the farmers to respond to their attacks. From behind a fortress of money-bagged barricades, the trust embarked on a war of attrition based on the age-old adage of "divide and conquer." Ultimately, the Amalgamated would spend between $600,000 and $1 million to defeat and discredit the claims of the association.[1]

The trust first initiated a campaign to undermine the key elements of the farmers' legal assault. John D. Ryan directed this part of the attack. Henry H. Rogers, the founder of Amalgamated, had first put Ryan in charge of conducting the "war of the Copper Kings" and driving his flamboyant counterpart, F. Augustus Heinze, out of Montana. One perceptive student of the episode considered Heinze to be "the most adept pirate in the history of American capitalist privateering."[2] If he was, then John D. Ryan, judging from the results, was his worthiest opponent. By 1906 Heinze was out of Montana copper, but that victory had required the implementation of perhaps the most unprincipled and lawless tactics ever employed in a major industrial conflict.[3] After finishing the fight against Heinze, Ryan, soon to be president of the Anaconda Copper Mining Company, turned to the Deer Lodge farmers as an incensed Gulliver might turn on the Lilliputians.

After the suit was filed and the farmers had expended forty thousand dollars, Ryan dispatched one Dr. O. Y. Warren of Butte to Idaho to find Fred J. Bliss and purchase his Deer Lodge ranch at whatever price he demanded. Bliss, however, could not sell because title to the land had been placed in the hands of attorney R. L. Clinton. Subsequently, lucrative offers were made to Clinton, allegedly by A. J. Shores, for the purchase of Bliss's property and for Clinton's removal from the association's case. Clinton refused the offers. Successive propositions included offers to purchase the farms of the association's leaders—such as Nick Bielenberg, W. C. Staton, chairman of the executive committee, and Conrad Kohrs—at very favorable prices. Staton saw the offers for what they were—an attempt to "terminate our interest in the fight," leave "100 farmers helpless," and cause them ultimately "the loss of their farms and homes."[4]

If the trust could not remove the source of the complaint, it could endeavor to cripple the prosecution of it by employing the time-honored expedient of bribery. Dr. D. E. Salmon, one of the most eminent of all the assembled experts and one of the few the farmers could bring in on their side, stayed during the trial at the Montana Hotel, the "enemy's camp." One sunny Sunday afternoon, prior to testifying, he was asked by Mr. Dobbins, manager of the hotel, to accompany him on a buggy ride through the countryside. Their ride proved so socially enjoyable that Dobbins invited Dr. Salmon to share a private box at the theater that evening. During the course of the evening's entertainment, Dobbins offered him first three thousand, then five thousand, and finally ten thousand dollars to desert the farmers. Dr. Salmon refused.[5]

Economic intimidation assumed forms other than protracted and expensive litigation. Plaintiff attorneys charged that hundreds of other farmers would have come forth to testify in the case, but "they were afraid it would jeopardize their interests and cause the enmity of the Company, which past experience has taught would be ruinous to position and property." Relatives of members of the farmers' association working at the smelter were fired—or given the "blue card." The superintendent of the smelter, A. J. Mathewson, did not "believe in furnishing powder for the enimie's [sic] guns." Dismissed workers were told to "Get your brother or relative to quit the Farmers' Association and you will be reinstated." "Only one conclusion will come to a careful, considerate mind," the farmers' attorney argued before the court of appeals in San Francisco: "They were afraid to be confronted with the truth." Why couldn't they try their case on its merits? he asked.

> It was not necessary for them to attempt to buy out Mr. Bliss nor to buy out Dr. Salmon. It was only with full knowledge that they were doing the injury and damage claimed to make them resort to such unlawful and subversive methods of justice and right that brought about these unlawful attempts to completely destroy the complainant's case. . . . Is not their whole defense *meretricious*? Is not this case one of the rank, glaring impositions which awakens anarchy and brings about loss of confidence and general criticism by those who know the facts? Is it not justly so?[6]

The "company press" was the most effective weapon, second only to the huge aggregate of capital, in the Amalgamated's arsenal. With the victory over Heinze in 1906, the Amalgamated acquired all of his mining interests plus his Butte newspaper, the *Reveille*. In the war of the Copper Kings, both Heinze and Ryan had used the press as a weapon. Amalgamated had constructed a whole complex of newspapers throughout the state to contend with Heinze for the public's opinion. When the battle was over, Amalgamated had a network of newspapers at its disposal. Those it did not own, it controlled by virtue of its overruling economic power.[7]

The Amalgamated, like all trusts, had its natural objectives. Its press championed and advanced these objectives. In the fight over smelter emissions, the objectives were at once specific and broad. In general, the press labored to shape the public's attitude toward accepting as natural law the preeminence of the trust's economic interests and priorities—to mold the community's mind to appreciating and accommodating itself to what Judge William H. Hunt defined dramatically as "the magnitude of the interest." Specifically, it concentrated on ridiculing the farmers and their claims of the destructive nature of the smelter's by-products. The *Standard*, now a part of the Amalgamated's empire, set the tone and worked for decades, at times subtly and at other times crudely, to make the concept of "magnitude of interest" the basic canon of Montana life.

Editorial comment and "news reports" extolling the productive capacity of the great plant and its meaning to citizens were regularly interspersed with or juxtaposed to reports of the Bliss trial. For instance, one morning in June 1906, Montanans opened the *Standard* to read: "Half a Million Dollars to Be Spent at Once on Additions to the Washoe Smelter Plant." The story under the headline explained the expansion was necessary to "balance the big plant on the hill." When construction was finished, 20 million pounds of copper would be produced monthly. "The plant," the paper explained, sharing its excitement with the reader, "already the largest in the world, will be a mammoth and its contribution to Montana's metal production will be enormous." In the event the connection between the welfare of the worker and businessman and the interests of Amalgamated was missed, it continued, "The increase in capacity means the employment of a good many more men at the smelters and a corresponding addition to the number of miners in Butte, contributing the ore supply to the Anaconda Plant. It is another demonstration of the progressive policy of the Amalgamated company's management and carries great significance to the cities of Anaconda and Butte."[8]

The people and the trust were obviously engaged in an enterprise serving humanity's ends. "Figures Tell No Lies," the *Standard* headlined in September 1906. "Facts pertaining to the direct and indirect effects of a closing down of the Washoe plant at Anaconda on the industries of the world which require copper were the main things brought out yesterday at the smoke hearings." Figures were cited to show that between 1902 and 1906 copper production increased nearly 100 percent. In "one single year the Washoe produced about one-fifth of the output of the U.S. and about 10 per cent of the entire output of the world," the paper reported. And "this production," it assured its readers, "is expected to increase." Therefore the court should, it argued, "consider the results that would follow the granting of a writ closing down the smelter and the effect on industries all over the world."[9]

The methods the farmers had employed were "drastic methods," the paper said. The "oft-expressed sentiment" during the progress of litigation "had been that all the involved communities want the farmers to get damages if they are found to be legally entitled thereto, but that none of these communities feel that everything they have should be put in peril or, indeed, swept away."[10]

On the date the appellate court in San Francisco rendered its decision on the appeal, almost six years from the date the suit was instituted, the *Standard* featured a half-page cartoon of an angry farmer taking a pratfall when the judge's decision cut the rope identified as the injunction strangling the smelter at the base of the stack. Citizens, miners, and businessmen were depicted cheering and throwing their hats in the air. Inserted in a cloud of smoke just puffed out of the top of the stack in three-quarter-inch type appeared the words, "Plenty, Peace, Good Times,

Cartoon from the Anaconda Standard, *March 7, 1911*

Prosperity." Below it in thick quarter type was a ludicrous anthem glorifying the smoke of the smelter:

> Oh say, can you see by the dawn's early light
> What so proudly we hailed at the twilight's last gleaming?
> Whose blue clouds and white mists through the long legal fight
> Over mountain and vale were so gallantly streaming!
> And the furnaces' glare, the whistles' keen blare
> Gave proof through the night that our smoke was still there!
> The smoke of the smelters! Long may it flow
> In the heavens above, o'er the earth here below![11]

From the beginning, the members of the farmers' association were labeled "smoke farmers"—too lazy to properly work the land, preferring instead to coerce money out of the smelter through threat of litigation. The smoke farmers had a "scheme," the *Standard* wrote in a long Sunday editorial titled "As It Is Written in the Chronicle of the Smoke":

> The scheme was one that beat the band. For it had come to pass that they had builded upon the high hill . . . a mighty smelter, a plant where the toilers obtained a red metal from the ore. . . . They responded with one another. And the words of Nick [Bielenberg] were heard in the land and they were the words that the people took unto their hearts, even as apples of gold and peaches. . . . Hearken unto me and hear. Listen that ye may profit and that ye add much gold, even shekels of fine gold and of silver, unto your store. And it will be a cinch. . . . And when they had heard these words they said, "Behold, is not Nick the real thing? Is it not so that we have no need to toil, neither to labor in our fields and to earn our bread by the sweat of our brows? For have we not a cinch and is it not a leadpipe?"[12]

The "chronicle" went on describing how the farmers were misrepresenting their claims and how the intelligence and wisdom of the officials of Amalgamated would liberate the community from such corruption.

Members of the farmers' association complained bitterly throughout the trial about the coverage and tactics of the press: "They are able to, and do, create a false understanding in all matters where it is to their interest to do so," said one leader of the association.[13] "No newspaper in the state seemed to have any information about the farmers' side of the case, but on the other hand the widest publicity was given to what the company chose to say concerning either side of the case," said another.[14] One paper, however, did choose to answer the association's direct query as to why their side of the story was not being reported. The *Powell County Call*, a Deer Lodge paper, gave a "frank statement of the Call's Attitude upon the 'Smoke Cases.'" It provided not only an expression of what probably most of the newspapers felt about the smoke cases but also revealed the concern with which most people viewed the idea of magnitude of interest.

The *Call* based its position on what it implied was social responsibility and what it said was "wise discretion sitting in the editorial chair." Its failure to raise a clamor was not "because of a lack of sympathy"; it was because invoking "public sentiment" now that the case was before the court would not assist the ranchers in obtaining "the fullest possible measure of just compensation for damages which are apparent and admitted." The *Call* chose instead, it said, "to make itself useful to them and to the people of the county as a whole by telling the story of Powell County's riches and resources, developed and undeveloped." The *Call* "looked to the bright side of things." Montana, as a result of "favorable advertising . . . [and] immense reclamation projects," of the "impending opening of rich and fertile Indian reservations," of low railroad rates and fabulous mineral riches was facing "a tide of immigration" that bade "fair to double the population of the state within half a decade of years." "In a word," the writer claimed, "there will be a newer and more prosperous and greater Montana!" The "issue," then, was whether the county should "receive its rightful share in this coming era of prosperity" or should "meet the incoming tide of investors and homeseekers with the red flag of warning, and crying out: Enter not! Within, death and desolation stalk in the valley!" The interests of the entire county, the article said, should not be jeopardized "over unfavorable conditions in a certain section, regrettable as those conditions are."

The *Call*, "weighing its words and willingly accepting full responsibility for them," concluded its frank statement on a note of astonishing self-deception to which the farmers could only shake their heads in despair: "In this, as in all matters of county concern, whenever it becomes apparent the *Call* may helpfully serve the people of this county, or any portion of them, the service will be rendered willingly and earnestly, and that without money and without price."[15]

While the Amalgamated attorneys were endeavoring to convince the court in Helena that the Deer Lodge valley had never been a rich farming region as claimed by the plaintiffs, the press was promoting it as an agricultural paradise and endeavoring to discredit the smoke farmers' claims that the smelter smoke was ruining the agricultural industry. In glowing accounts, the *Standard*, the *Inter Mountain*, and other papers told of the profundity and quality of the valley's products. Officials of the Anaconda Copper Mining Company took over the organization of the Deer Lodge fairs. Superintendent Mathewson even took time off from his metallurgical responsibilities at the Washoe to direct the annual event.[16]

The ballyhoo began in 1905 at approximately the time testimony in the Bliss case was getting underway. No device was too absurd, nothing too transparent. Perhaps the best example of the lengths to which the press would go was a description of the valley's showing in the state fair in Helena, printed in the *Inter Mountain*. The Deer Lodge agricultural display "equals in quality that of any county competing for prize," the paper said. "No collective exhibit in the building is attracting more attention than that of Deer Lodge. . . . It is because many persons formed

Deer Lodge exhibit at the Montana State Fair, September 25, 1905. The Anaconda Company promoted such exhibits to counteract farmers' claims that smelter smoke was ruining the Deer Lodge valley's agricultural potential.
MONTANA HISTORICAL SOCIETY PHOTO ARCHIVES, HELENA

the idea that Deer Lodge county has been ruined from an agricultural standpoint by smelter smoke." The correspondent felt compelled to point out that the "exhibit is entirely representative, for no effort was made to secure the best of its product." Only the "average sample" of each thing grown in the valley was shown, he said. Under the title, "It Surprised Him," a dialogue between the director of Deer Lodge's exhibit and a farmer from eastern Montana was carefully recorded:

> A Yellowstone farmer ambled up and looked with admiring eyes upon a collection of turnips, each about as big around as a bucket, which formed a part of the exhibit.

> "And so this is the Deer Lodge exhibit?" he inquired.
>
> "Yes sir," said Mr. Felix Greenan, who is always pleased to meet visitors.
>
> "And these things were grown near Anaconda?"
>
> "Yes sir, within a few miles of the town."
>
> "Well, that's in smoke country, isn't it?"
>
> "That's what it is called," replied Greenan smilingly.
>
> "I guess you people can raise something over there besides smoke," exclaimed the Yellowstone man, who walked away.[17]

The *Standard* pronounced the Deer Lodge exhibit at the 1907 state fair the "leader." "In the Deer Lodge exhibit," it said, "the Anaconda Copper Mining Company has played a prominent part." It explained that "General Manager E. P. [sic] Mathewson has taken a personal pride in the showing to be made, and the results are creditable both to himself and the county." As was the case in previous fairs, "every variety of farm product that can be grown in the county [is shown], and while the variety is great," the paper emphasized, "the quality is what impresses."[18]

As the trial approached its conclusion in Helena, the ballyhoo intensified. Repeatedly Montanans were told in headlines and subtitles that there were "Good Farms in Deer Lodge Valley," there were "Opportunities in Deer Lodge Valley," farms of "Small Acreage Well Cared for Will Give Good Returns," the "Markets Are Unexcelled," and "Big mercantile establishments were prepared to furnish the farmer everything he wants located in near-by towns." Coincidentally, one contention of the smelter's attorney was that the members of the farmers' association were practicing poor farming methods. The production of wild hay, the *Standard* pointed out, was constantly on the increase. "Where careful attention was paid to conditions," it explained, as much as five tons of alfalfa to the acre could be cut in a season.[19]

If the farmers agreed with nothing else in the *Standard*, they did agree with its statement that the Anaconda Copper Mining Company was playing a "prominent part" in the Deer Lodge fairs. Too prominent. "The defendants have been conducting a county fair in Deer Lodge County since the farmers have been complaining of damages, and there is not a farmer or land owner in the Valley who is an officer, and they consist of the smelter management and merchants of Anaconda," complained Walter Staton, chairman of the executive committee of the farmers' association. "The yield of our hay land are [sic] steadily decreasing," he asserted. In 1906, first-prize wheat went fifteen bushels to the acre, he claimed. But in 1907, the farmer who won the prize, Frank Kinney, who lived west of Anaconda, "did not thresh at all" because there was not "sufficient wheat to warrent [sic] threshing."[20]

The Anaconda Company had established "experimental farms" at three points forming a triangle in the valley. On the farms they raised crops and livestock. They also claimed to be raising pure-blood Herefords and entered the cattle in various county fairs and in the state fair as "Deer Lodge cattle." But the farmers claimed the herd had been purchased in 1907 and kept in "close domestication" with a well-paid "special man" to care for them. One herd consisted of fifteen head, said the farmers, mostly bulls, but since 1907 they had produced only two calves. The calves, the farmers calculated, cost the Anaconda Company six hundred dollars each to produce, "and doubtless," they observed dryly, "they have used their splendid success with these cattle . . . as an argument here."[21]

The farmers protested the makeup of the board of the Deer Lodge fair and presented affidavits to the board claiming the entries by the experimental farms were "false and fraudulent." The board rejected the objections and the farce continued. The fair was "absolutely controlled by the Smelting Company," the farmers complained in vain.[22]

While the press was painting a picture of the valley as a thriving agricultural region requiring only "careful attention" to turn a profit, testimony at the trial was developing a contrary view. When the company tried to market twenty prize steers from one experimental farm, eight were rejected outright and the remaining twelve were sold at greatly reduced prices to Montgomery & Company—a packinghouse owned by the Anaconda Company. Another fifteen steers pastured on the Bliss ranch shared a similar experience. The smelter's attorneys allowed for only one examination to be made of the stock and entered into evidence. The examination was made by Dr. Robert J. Formad of the U.S. Bureau of Animal Industry. He selected one of the best steers on the Bliss property; his autopsy of the stomach and various organs showed characteristic lesions of arsenical poisoning. The lesions were of the same type as lesions found in an animal the plaintiffs had experimented upon by feeding it controlled doses of arsenic. The only difference was that the lesions in the steer Formad chose were not yet as marked and acute as those in the cow deliberately poisoned.[23]

Grain farming was also carried on at the experimental farms and the results were similar. Attorney Clinton claimed that a threshing machine had never "thresh[ed] a single straw of grain" on any of the company's farms. He repeatedly challenged them to introduce the results of their farming into the court record to show the nature of the operation as to cost, crop, and the profit or loss. "The record will show," Clinton said, "that they promised from time to time . . . to produce these results, but they defaulted." This, he concluded, left undisputed the farmers' testimony that farming or stock raising could not be carried on at a profit in that portion of the valley subject to smelter emissions.[24]

A group of scientists making an investigation of the country surrounding the smelter for the U.S. Department of Justice in 1910, after the trial court's decision, observed the company farms and reported that they "looked exceptionally green and flourishing from the road, but on examination proved to be mainly in either potatoes, which are fairly resistant to smelter smoke . . . or grain, which is excellent as to vegetation, but scant as to grains in the heads." Their fields were also "heavily fertilized and well irrigated," the investigators said.[25]

No assistance was available to the embattled farmers from the Montana congressional delegation. Indeed, they were another source of adversity. The state's two most influential congressmen during the duration of the trial were Sens. Thomas H. Carter and Joseph M. Dixon. Senator Carter was a shallow man, representative of that class of politicians who conceive of politics as principally the conduct of public affairs for personal advantage and the advancement of the objectives of special interests. In Carter's case, the concern for special interest was almost exclusively confined to the Amalgamated. In reading his papers, one can only with great effort escape the conclusion that he spent most of his congressional efforts escorting Amalgamated officials around Washington, D.C., and traveling hurriedly between their offices in New York and the Capitol.[26]

During the Bliss trial, Senator Carter sponsored a piece of legislation that almost crippled the embattled farmers. In 1908, a year before Judge Hunt's decision, he added an amendment to the Sundry Civil Appropriations bill of the Sixtieth Congress, doubling the fees and compensation of the clerk of the circuit and district courts for the District of Montana. The bill became law and increased the price of a transcript per sheet on appeal from the circuit court to the district court from fifteen cents to thirty cents. As a result, when the farmers moved to have Judge Hunt's decision appealed, the cost of appeal was double what it would have been. The farmers' association, already bankrupt, saw this as another underhanded practice of the Amalgamated. "We take it," said a spokesman, "that this bill was introduced and pushed to its passage to prevent the possibility of an appeal on the part of the farmers of Deer Lodge Valley in their litigation with the smelter."[27]

Senator Dixon was the antithesis of Senator Carter. Joseph Dixon was no toady of the trust. He combined a fine balance of ambition, idealism, and political savoir faire. Yet in the farmers' case, the practical realities of politics barred him from doing anything more than recommending that they sue for damages and expressing the sincere hope that they get their "just dues." This afforded little comfort to the frustrated farmers, who saw their position thusly:

> Senator Dixon states: "Let the farmers bring damage suits." If the farmer could show every cent of damages sustained (which is impossible), and could get a verdict in court for every cent to it, the lawyers fees, evidently, would leave him bankrupt, as he would have to sue every crop season and for every horse and cow that died; and to prove the death of one horse by expert testimony would cost him more than his entire herd is worth.[28]

On November 14, 1908, months before Judge Hunt's decision, the farmers' worst suspicions were confirmed—justice was not blind. The newspapers informed them that Judge Hunt had been entertained in jovial good fellowship at "an elaborate affair" at the home of Judge John J. McHatton in Butte. Assembled around the dinner table decorated with red carnations and smilax, which "presented a very pretty picture," were a dozen of the most prominent citizens of Butte, and conspicuous among them were John D. Ryan and Cornelius F. Kelley. A farmer from Waldheim, Montana, correctly concluded that the "wining and dining" of Judge Hunt pending a decision on the case was arrogance and "oppression by the rich" of the first order.[29]

In 1908 the association, having endured social ostracism at the hands of the press, sensing an adverse decision at the hands of a biased judge, and facing bankruptcy, turned to the most prominent man in the land for succor. In late 1908 they began a series of letters to Pres. Theodore Roosevelt telling of their plight and of the damage to their land. The decision to turn to the chief executive was a logical one. Theodore Roosevelt considered himself a "western man." And Conrad Kohrs, the most prominent man in the Deer Lodge Valley Farmers' Association, was a close friend of Roosevelt's. They had become acquainted years before when they were both members of the Montana Stock Growers' Association. Since that time, Kohrs had become a successful cattleman and was locally prominent in the Republican Party. He and the president corresponded fairly regularly, usually discussing political strategy.[30]

The example of an association of farmers contending with a powerful industrial combination had great attraction for Theodore Roosevelt. It affirmed a concept he had articulated that very year: "The chief breakdown [of the American political system] is in dealing with the new relations that arise from the mutualism, the interdependence of our time," Roosevelt told Congress in 1909 in his last annual message to that body. "Every new social relation begets a new type of wrongdoing—of sin, to use an old-fashioned word—and many years always elapse before society is able to turn this sin into crime which can be effectively punished at law."[31]

Through "mutualism," Roosevelt hoped to solve this problem. He hoped to create a political environment conducive to social and economic combinations he thought the nation required. Government, through responsible administrative agencies, would prescribe the rules by which these various combinations would operate. Eventually, government would preside, Roosevelt believed, over an "equilibrium of consolidated interests" composed of agriculture, labor, and industry. As this doctrine applied to agriculture, Roosevelt proposed:

> Farmers must learn the vital need of cooperation with one another. Next to this comes cooperation with the government, and the

> government can best give its aid through associations of farmers rather than through the individual farmer. . . . It is greatly to be wished . . . that associations of farmers could be organized, primarily for business purposes, but also with social ends in view. . . . The people of our farming regions must be able to combine among themselves, as the most efficient means of protecting their industry from the highly organized interests which now surround them on every side.[32]

The appeals of the Deer Lodge Valley Farmers' Association struck an unusually responsive chord in Roosevelt and his administration. In addition, and perhaps most importantly, the farmers were wise enough to incorporate in their pleadings for assistance the powerful argument of conservation. Conservation of natural resources was to Roosevelt more than a means of preserving a domain for agriculture or industry; it was a means of preserving and enhancing the strength of the whole nation.

Two letters from the association to Roosevelt ran thirty typewritten pages. One was under the heading "Why the Farmer Should Have the Protection of the Government." In Montana, the chair of the association's executive committee told Roosevelt, "the smelters seem to think 'might is right.'" "We would not need to be here asking the assistance of the general Government if our officers of the state would act and protect the citizens," Walter Staton asserted. The association had attempted to arbitrate its differences with Amalgamated, he said, through the lieutenant governor, even agreeing to bind itself to dismiss all actions against the trust. The offer was refused. The State was impotent to act even though the conduct of Amalgamated was unconstitutional. The state constitution—"framed before Amalgamated control," Staton inserted pointedly—specified that "No corporation shall so conduct its business as to infringe upon the equal rights of an individual."[33]

The cost of the suit to date had been $200,000 and the cost of an appeal, for the transcript alone, would amount to $50,000. What was worse, Staton declared, was that five seasons had passed "on which not one cent of damages [has been] paid to date." The cost of litigation was "more than the best farm in the Valley" was worth, he said. "And this the company knows," he asserted. Many farmers had gone bankrupt, and many more would follow. The banks in Anaconda would not lend money on "smoked land." They would not "consider farm lands under present smelter conditions as security." The costs were described in concrete terms. The damages and experiences of dozens of farmers were detailed, including those of Roosevelt's friend, Conrad Kohrs. One farmer-rancher who owned three thousand acres and who, previous to the opening of the Washoe smelter had earned eight thousand dollars a year, was now "making a living for himself and family by working on the county road as road supervisor."[34]

The core of the farmers' argument to Roosevelt and the most telling, they thought, was that they were pioneers. A great many of the farmers had U.S. land patents issued and signed by Pres. Ulysses S. Grant, and "some were Civil War veterans homesteading under soldiers' homesteads." "Having gone through the privations and danger of first pioneers and receiving their titles direct from the United States Government," the petition asked, "are they now not, under the terms of the grants, entitled to protection from the government? They are too old to start anew and unable to make a living under the injurious affects [sic] of the smelter fumes. . . . These pioneers did not come to the smelter, but the smelter came to them."[35]

"We were compelled to combine and sue for an injunction as the only means of redress," the farmers wrote Roosevelt. But Amalgamated had "such political power in Montana that they can secure any kind of written statements, and through fear of financial loss, the business interest follows suit." "[We] offer arbitration," Staton said, but "the company offers endless litigation." The smelting company's "main defense is the magnitude of their operations," which, the farmers declared, "is not just defense." The Amalgamated "pleads the magnitude of their operations in order to escape the consequences of their unlawful and predatory acts." What the farmers wanted, Staton told the president, was "relief, either by the purchase of their holdings or the poison removed from the fumes."[36]

The association's pleas ended with an emphasis on two ideas that were as canon law to the patrician-minded Roosevelt: duty and conservation. Concluding that the Washoe smelter must be forced to control its emissions, the farmers said,

> Now, Mr. President, we believe there is a duty on the part of the Government to protect its woods and watersheds in these National Forest Reserves, and if this protection is delayed for any long period the injury will be complete and protection will come too late. We, therefore, respectfully urge that the Government take immediate action in court to protect these interests, which will also be a direct protection to the farming interests herein called to your attention.[37]

– Chapter Eight –

The Roosevelt Men versus the Smelters

Theodore Roosevelt, by virtue of intellect, experience, and philosophy, was peculiarly equipped to undertake the task of forcing the Amalgamated to control its emissions at the Washoe smelter in order to protect the national forests and indirectly aid the farmers of the Deer Lodge valley. Behind the enthusiastic smile and the glittering spectacles dwelled a sophisticated intelligence. The man had a shrewd understanding of American politics that few people of his time could match. The approaching confrontation represented more than a dispute over the emissions of the Washoe smelter. In a larger context it was a clash between two theories of economic and political organization: laissez-faire on the one hand and the general-welfare state on the other.[1]

Laissez-faire had become a reality in the 1840s and 1850s, before the development of the trusts had constricted free competition and centralized the economy. By the end of the nineteenth century while its viability had faded, its popularity as an "idea system" was paradoxically at its highest. The idea of the general-welfare state emerged in America as the reality of laissez-faire was disappearing. The proponents of the former argued that the state could best promote the general welfare by a positive exertion of its powers. Laissez-faire advocates, on the other hand, had proposed that the government could best advance the commonwealth by rendering itself as inconspicuous as possible and by permitting individuals to work out their own destinies amid the free and unrestricted play of natural forces. This was the theory, at least in the minds of the social Darwinists and political economists. In practice, especially in the two decades preceding the presidency of Theodore Roosevelt, laissez-faire had become a rationalization of the status quo, particularly in the hands of industrialists. While allowing for subsidies, protection, and favors from the government for business interests, the doctrine at the same time afforded a means of discouraging conscious efforts at reform by government, particularly on behalf of labor and agriculture.[2]

Theodore Roosevelt, as president, became the foremost activist on behalf of positive government and the leading opponent of the plutocratic trend. Writing to a friend, Seth Low, in 1908, he asserted: "What I want is more power for the Government to decide by executive action what is and what is not proper. . . . [s]o as to prevent the gross abuses that have existed and that exist." The government had to look out for the interests of all the people, not just a privileged few. "The war we wage," he lectured Congress in the same year, "must by waged against misconduct, against wrong-doing wherever it is found; and we must stand heartily for the rights of every decent man, whether he be a man of great wealth or a man who earns his livelihood as a wageworker or a tiller of the soil." Roosevelt, then, aimed to govern. This was the point of conflict with his opponents, for the leaders of American industry and finance, of which Amalgamated officials were representative, also aimed to govern.[3]

Two of Roosevelt's goals as president were to elevate his office as the dominant force in the national government and to make that government, while always acting in the public interest, the most important single influence in national affairs. From his opponents it earned for him the persistent charge of being a dictator. But Congress was not equipped to evaluate various demands, set the required priorities, vote on them swiftly, and initiate their administration. The executive branch, Roosevelt perceived, was in a unique position to fulfill these very functions. Roosevelt's idealism and charisma drew into his cabinet a group of unusually competent, dedicated, and ambitious men, willing to assist him energetically in expanding the influence of the executive. They were extraordinarily adept at the art of politics, and as a result of sharing an unusually deep fellowship and common philosophy, they formulated in a very short period a policy for the regulation and use of natural resources. All of the president's advisors understood and appreciated that their policy of conservation had to be exercised within the limits of traditional politics and with popular approbation.[4]

The "fundamental evil" in a democracy, the conservationists held, was the concentration of wealth in the hands of the few, and the basis for the concentration of wealth was the monopoly of resources. "Conservation," they argued, "is the most effective weapon against monopoly of natural resources." The nation had been using up its natural resources at an alarmingly accelerating rate, they reasoned. Destruction and exhaustion of these resources appeared imminent. To avert the disaster, the farsighted Roosevelt gave the conservation movement his blessings from the time of his first message to Congress, and the cause flourished. Roosevelt, however, was always cautious never to get too far ahead of current opinion or too far outside of customary values for fear of losing the power to lead. He therefore refined and articulated his position on conservation to fit the attitudes of responsible, middle-class Americans. "The fundamental idea of forestry," he said in his first message, "is the perpetuation of forests by use" and the search for a means of increasing and sustaining resources and the industries dependent upon them. "The preservation of our forests is an imperative business necessity," he said.[5]

By 1908 the condition of the forests was of paramount concern to Roosevelt's conservation program.[6] A major portion of his Eighth Annual Message to Congress was devoted to the theme of saving the forests and thus preserving the nation. Drawing upon the history of the destruction of the forests in China, North Africa, and Europe and the consequences of that destruction to the well-being of these areas, he pointed to small sections throughout the United States—in the Adirondacks, the White Mountains, the Appalachians, and the Rocky Mountains—that bore permanent scars from Americans' efforts to subdue the country. Warning that continued "wreckless deforestation" was an invitation to disaster, Roosevelt put forth in his typical rousing and didactic style the imperatives of "the Roosevelt men":

> We should leave our national domain to our children, increased in value and not worn out. . . . If there is any one duty which more than another, we owe it to our children and our children's children to perform at once, it is to save the forests of this country, for they constitute the first and most important element in the conservation of the natural resources.
>
> Nothing should be permitted to stand in the way of the preservation of the forests, and it is criminal to permit individuals to purchase a little gain for themselves through the destruction of forests when this destruction is fatal to the well-being of the whole country in the future.
>
> Short-sighted persons, or persons blinded to the future by desire to make money in every way out of the present, sometimes speak as if no great damage would be done by the reckless destruction of our forests. It is difficult to have patience with the arguments of these persons. Thanks to our own recklessness in the use of our splendid forests, we have already crossed the verge of a timber famine in this country, and no measures that we now take can, at least for many years, undo the mischief that has already been done. But we can prevent further mischief being done; and it would be in the highest

> degree reprehensible to let any consideration of temporary convenience or temporary cost interfere with such action, especially as regards the national forests which the nation can *now*, at this very moment, control.
>
> In the total absence of regulation of the matter in the interest of the people, each small group is inevitably pushed into a policy of destruction which cannot afford to take thought for the morrow. This is just one of those matters which it is fatal to leave to unsupervised individual control. The forest can only be protected by the State, by the nation; and the liberty of action of individuals must be conditioned upon what the State or nation determines to be necessary for the common safety.[7]

With the heightened anxiety for the preservation of the national forests and the concurrent receipt by the Department of Agriculture of widespread complaints about destruction of forests and crops by smelters, the department, consistent with the ideas of positive government, initiated a series of investigations to ascertain the facts and sponsored a number of studies looking toward a resolution of the problem. Thus, control of smelter emissions became an integral part of the Roosevelt administration's conservation program.[8]

The areas of major complaint were in Tennessee, Utah, California, and Montana. Considerable damage had been done at Mackay, Idaho, by the operations of the White Knob Copper Company. As yet, Wyoming was relatively free from smelter damage except in the area around Encampment. In the cities of Pueblo, Leadville, Durango, and Denver, Colorado, and in Tacoma and Seattle, Washington, there were considerable complaints of fume damage by smelters situated within their environs. But with the exception of California and Montana, little or none of the public domain was situated within the smoke belt of the offending smelters. The federal government could only become directly involved in those areas where public lands, principally forest reserves, were either undergoing substantial damage or were in jeopardy. From the Roosevelt administration through the Taft and Wilson administrations, attorneys in the Department of Justice patiently and apologetically explained to citizens complaining of smelter fumes that the U.S. government could not litigate for the benefit of one citizen against another. Consequently, the involvement of the federal government was limited to northern California and western Montana, where national forest reserves were situated. Nevertheless, studies on the effects of smelter emissions on vegetation were carried out in all regions where complaints were received.[9]

As early as 1903, J. K. Haywood, chief chemist of the Miscellaneous Division of the Bureau of Chemistry, was in northern California, studying the effects of sulphur dioxide and sulphur trioxide on the vegetation surrounding the Keswick smelter near Redding. The study extended through 1904 and was published in 1905. Among his conclusions, Haywood found that sulphur dioxide "when present in very minute amounts in the air kills vegetation," and vegetation around the smelter, as far away as nine miles, was "greatly injured." There was "less severe injury" for a considerable distance beyond. Photographs of the damage to trees and vegetation accompanied the study. In Tennessee, through the years 1905 and 1906, Haywood made similar studies near the cities of Ducktown and Copperhill and drew similar conclusions. He found, though, that the damage encompassed a larger area in Tennessee than in California.[10]

At Anaconda in 1906 and 1907, Haywood, assisted by a forester, conducted the most comprehensive investigation to date. He sought not only to determine the effects of sulphur dioxide and trioxide on vegetation but also the amount of arsenic emitted from the Washoe smelter and the effect of the arsenic on livestock. He also sought to discover whether the waste discharged from the smelter into the streams rendered the water unfit for irrigation and whether it damaged the soil. Supporting his results with photographs, Haywood drew several "definite conclusions." On the basis of "actual chemical analysis," Haywood's conclusions were directly opposite those of Judge William H. Hunt in the Bliss case, which was simultaneously drawing to a conclusion. The Deerlodge National Forest and other forests on public land had been injured for at least ten miles north of the smelter in line with the farms of the Deer Lodge Valley Farmers' Association, for at least six miles south of the smelter, and for at least thirteen miles west of the smelter. The injury "undoubtedly" extended even farther, Haywood concluded—as much as ten miles farther north and at least two more miles south and east of the smelter.[11]

To the east of the Washoe, toward Butte, the damage extended fourteen miles and the damage from the Butte smelters overlapped the damage from the Washoe. The principal damaging agent to trees was found to be sulphur dioxide. The trees almost impervious to the smelter fumes were junipers; the most susceptible were red firs, which were "badly damaged." The lodgepole pines were more resistant than the red firs but showed damage for at least ten miles from the smelter. Large quantities of arsenic, as much as sixty-six tons per day, were discharged either in the form of fumes or in the tailings and slag that emptied into the Deer Lodge River. The amount of arsenic found in forage crops within ten miles of the smelter was enough to poison cattle. "In order that the cattle in this region may live at all," Haywood observed, "it is evident that they must become confirmed arsenic eaters." Furthermore, the Deer Lodge River had been rendered unfit for irrigation as a result of the smelter's discharges. Land irrigated was "greatly injured by the copper present in the irrigation water." Finally, Haywood reported, there were not enough alkali salts in the soil of the Deer Lodge valley to injure ordinary farm crops.[12]

The most significant finding from Haywood's investigations of 1906 and 1907, the district forester in Missoula concluded years later when reviewing the studies, was that damaged trees within the area affected by smelter fumes showed sulphur dioxide and sulphur trioxide present in "definite quantities," while trees in districts outside the range of smelter fumes contained only a trace of sulphur in the leaves. Throughout the same period, other investigations seeking to establish the cause or causes and extent of the damage to the forests had gone far, the district forester concluded, "in eliminating the possibility of the damage being done by insects, fungi, fire, and weather."[13]

In 1906 the administration, under the direction of Attorney General Charles J. Bonaparte, a patrician reformer with a well-developed sense of noblesse oblige, determined to move against the smelters to prevent continuing injury to the national forests and public domain. The first prosecution was against the Mountain Copper Company of California. The plant was situated on San Pablo Bay near San Francisco. The government was successful in securing an injunction and closing the plant because of fume damage to an adjoining public forest. As many as six thousand acres of forest had been killed or damaged by the fumes. On appeal, the injunction was overturned. The federal government appealed that decision to the U.S. Supreme Court, but partly because of limited resources and partly because of an incompetent attorney, the record in the case was in such "an exceedingly defective condition" that the government was afraid to go to trial before the Supreme Court. An adverse decision, they reasoned, would establish an "exceedingly harmful precedent against the government" and subject the United States to enormous and widespread injury.[14]

To straighten out the government's case and "fully preserve the government's interest," the Justice Department secured the services of attorney Ligon Johnson of Atlanta, Georgia. A committed progressive and conservationist in the mold of his boss, Attorney General Bonaparte, Ligon Johnson was destined to become one of the foremost "smoke

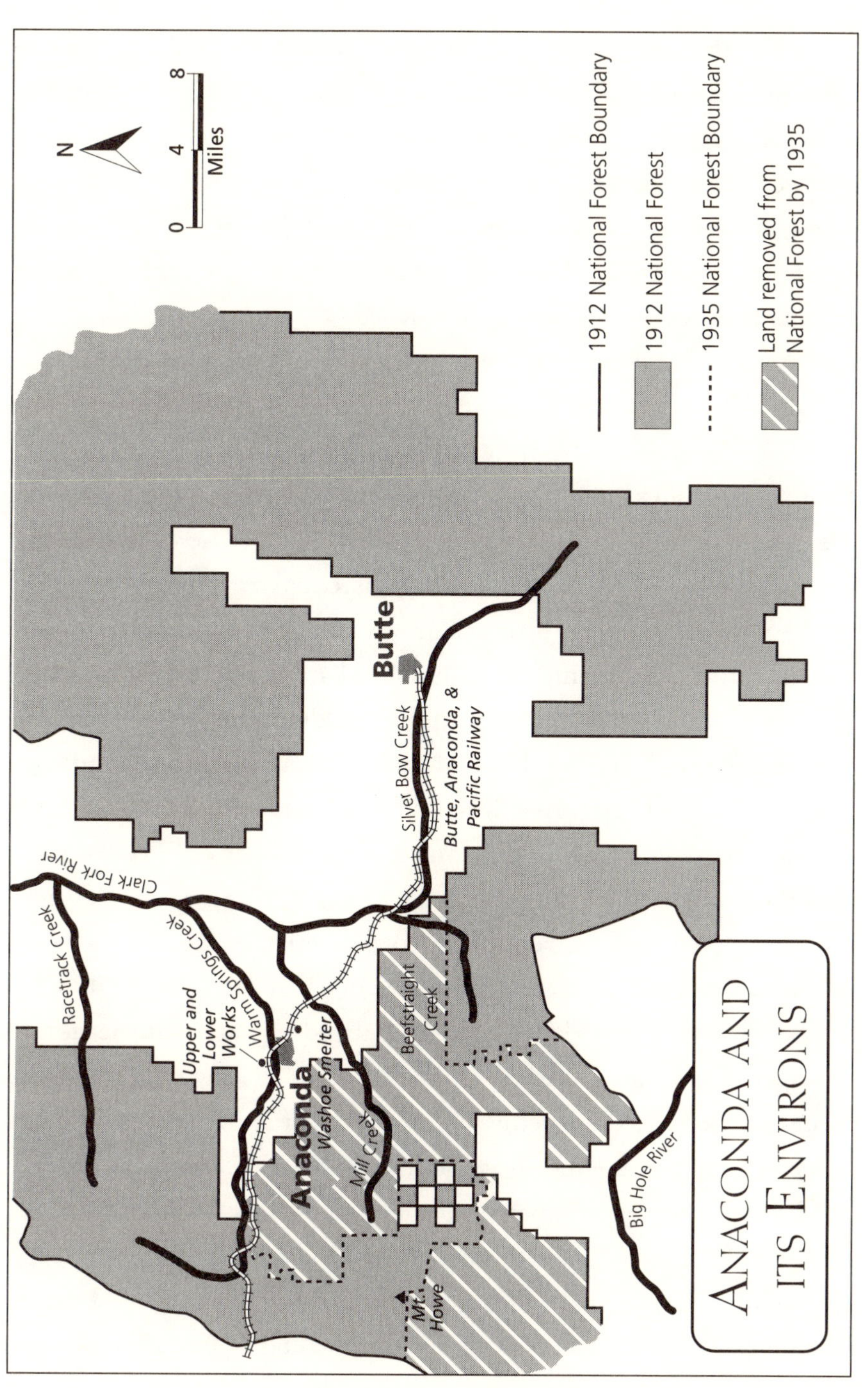
Anaconda and its Environs
N
0 4 8
Miles
1912 National Forest Boundary
1912 National Forest
1935 National Forest Boundary
Land removed from National Forest by 1935
Butte
Anaconda
Washoe Smelter
Upper and Lower Works
Clark Fork River
Racetrack Creek
Warm Springs Creek
Silver Bow Creek
Butte, Anaconda, & Pacific Railway
Beefstraight Creek
Mill Creek
Mt. Howe
Big Hole River

counsels" in the country. The department was fortunate in getting Johnson. He was employed at the time as counsel for the State of Georgia in its suit against the Tennessee and Ducktown Copper Companies, which involved almost the same law and conditions involved in the government's case against the Mountain Copper Company. Johnson had just won the case for Georgia against Tennessee and Ducktown Copper after a number of suits against them by farmers had failed.[15]

In the Tennessee case, crops of farmers and property of the State of Georgia were being destroyed by the fumes from the Tennessee Copper Company at Copperhill and by the Ducktown Copper Company at Ducktown. The smelters were situated near the Tennessee-Georgia state line and their fumes drifted south into Georgia. The litigation involved two separate suits prosecuted by Johnson on behalf of the State of Georgia. The first suit was instituted in 1904 to stop the smelters from eliminating the sulphur and arsenic by heap roasting—the same process utilized in Butte a decade and a half earlier and terminated only after the threat of mob action. The area surrounding the smelters had been turned into a barren desert. During the course of the suit, representatives of the companies and their experts proposed that the problem could be remedied by a change in smelting methods. They proposed the construction of a plant similar to the one "in vogue at Anaconda," where the smelting was done in furnaces "with a tall stack to carry the gases high in the air," affording natural dispersion of the gases into the atmosphere. The State of Georgia accepted the representations of the smelter's experts as a "true remedy," and upon agreement by the smelters to abandon heap roasting and install the new method, the suit was dismissed without prejudice.[16]

Predictably, the "natural way" failed. "Operation of the new method of smelting," the State of Georgia soon learned, resulted in "damages vastly greater than under the old heap-roast method." Once again Georgia filed suit, asking that the smelters be enjoined from operating until the poisonous substances were removed from their emissions. The case was fought through the various stages of demurrers, pleas, and temporary restraining orders until the State of Georgia was finally granted a permanent injunction. The lesson was clear to Johnson. Coercion was the only effective weapon. Writing to Attorney General Bonaparte, he observed, "Not until Georgia was given this powerful weapon, did the companies take steps toward the manufacture of sulphuric acid [as a means of removing sulphur from the fumes] and it was after this decision had been rendered that the companies agreed, if the injunction was suspended, that they would immediately begin the installation of acid plants. All this is borne out in detail by the records of the case in the Supreme Court of the United States."[17] Thereafter, the use of injunctions became the modus operandi of the administration in contending with smelters that damaged public lands.

Months before hiring Johnson, Attorney General Bonaparte had been contemplating the institution of a number of suits against the smelters in California and Montana to force them to abate the poisonous substances in their emissions. But, in a fashion typical of the Roosevelt administration, he was reluctant to proceed until he was confident he could win.[18] Consequently, while Johnson was in the West attempting to remedy the Mountain Copper Company case he was instructed to ascertain the damage to public lands, and the legal aspects pertaining to such damage, and to reconnoiter the region and sound out the smelter industry in terms of the political and economic ramifications a suit might force the administration to contend with.

Upon returning to Washington late in November 1907, Johnson informed the Justice Department that efforts to obtain the best adjustment possible in the Mountain Copper case were underway. As a result of increasing copper production, a corresponding increase in fume injury to the national forest and public domain had occurred. Damage was "fairly widespread and growing rapidly," Johnson reported. As a result of incineration of the ores, deadly gases and wastes were released into the atmosphere and into streams, all of which was "highly injurious" to animal and vegetable life. When these streams were used for irrigation, Johnson wrote, the fertility of the soil was destroyed and "plant reproduction or even plant life" was jeopardized.[19]

The areas of greatest damage were in Montana and California. In California the "chief fume injuries" were confined to the smelting plants of the Mountain Copper Company at Keswick and Kennett and the plant of the Bully Hill Copper Mining and Smelting Company at Del Mar. The plants at Keswick and Kennet had combined to make "practically a continuous desert strip" of the Sacramento canyon, and the damage reached government lands. The Bully Hill Company was constructing extensive facilities to increase its production, so the damage could be expected to "increase enormously in the future," Johnson warned.[20]

The consequences were potentially ruinous:

> The California smelters are particularly disastrous to vegetation and the water supply generally by reason of the fact that they are situated in one of the sections having the heaviest rainfalls in America. This results not only in greater injury to vegetation, by reason of the gases being held close to the earth and more constantly precipitated upon as well as breathed in by the surrounding plant life, but also of injury by floods and freshets . . . soil erosion, loss of water supply and the like.

There was, however, a bright side to the California situation, Johnson reported, apparently thinking in terms of remedies: "The ore here has no arsenic contents."[21]

Johnson found that the worst destruction in the country was taking place in Montana. "At Anaconda, the damage is extensive and far-reaching, probably covering a thousand square miles of territory," he estimated. As a result of the heavy arsenic content in the ore, cattle feeding upon the vegetation were poisoned, "even at considerable distance from the smelter." The smelter emissions affected the Deerlodge National Forest as well as the public domain. Ultimately, an effect on the water supply would be felt, he said, because the tailings from the smelter had already poisoned the irrigation canals.[22]

The same situation prevailed in the Butte region. "The vegetation has already been destroyed . . . and the effects reach both the national forests and the public domain," Johnson reported. Even so, the harm was "not so marked as at Anaconda." At Great Falls the damage radiated for a number of miles around the plant, but "the results are not as disastrous as in the mountainous regions of Anaconda and Butte." The topography at Great Falls was fairly level, composed of high tablelands with no forests. The vegetation consisted mainly of grass, and while the sulphur fumes had rendered the area unfit for grazing, "the level character of the land prevents soil erosion." Consequently, the water supply was not as endangered as in the "more rugged and declivitous localities."[23]

The political and economic situation, Johnson explained at length, was very complex. In the context of the proposed litigation against the smelter interests, he cautioned, it was politically explosive and had to be handled judiciously. His investigations had convinced him that there had been an overproduction of copper and that as much as 300 million pounds of copper were in the hands of the United Metal Selling Company, which it was unable to market. Through the medium of United Metals, Johnson explained, the Amalgamated and allied interests had doubled the price of copper in the last three years, placing the cost of the metal at "practically prohibitive figures."

Consumers had been forced out of the market, yet production had "steadily forged ahead with the result that enormous supplies" were now on hand. When American producers had tried to stimulate consumption and meet the selling price of foreign producers who had been undercutting the American combination, the price of copper had tumbled from twenty-six cents a pound to sixteen cents. "Instead of renewing the normal consumption of copper and confidence of the trade," Johnson said, "the situation is as badly disorganized as ever as few or no sales are being made." As a result of this poor market, he concluded, "All the large mining interests of the West would be more than willing to close down temporarily and the Amalgamated at Butte, Anaconda and elsewhere has already prepared to curtail operations and possibly to suspend temporarily." And this, he stressed, was where the danger lay:

> A suit at the present time would more than likely be seized upon as a pretext to close and let the cry go out that under the persecution of the administration thousands of miners had been thrown out of work. The copper interests and Standard Oil are, of course, almost identical, and I think there is little doubt that a suit against the Amalgamated at Anaconda would result in a general closing down until a considerable amount of the stored copper could be marketed.

The shutdown would then accrue to the benefit of the copper trust and to the detriment of the Roosevelt administration. "An additional result to be expected," Johnson reasoned, "is an increase of copper prices as a suit by the United States would not only steady the market but give copper an upward tendency leaving copper producers in a position to profit by the situation financially and malign the administration at the same time."

Much of Johnson's assessment of the intentions of the copper industry proved correct. Within a short time after his first report, several of the smelters closed and the Amalgamated curtailed its production. "Fortunately," Johnson pointed out to Attorney General Bonaparte, "there were no suits pending to be seized upon as a pretext of explanation for the failure of United Copper, the laying off of miners and employees and general financial troubles."[24]

Johnson gave his personal recommendations as to how the administration should proceed with the case. He thought it "wiser to do nothing now beyond disposing of the Mountain Copper Company case" and to use the coming winter, spring, and summer for the "full preparation of a case to form a precedent." Then in fall 1908, when the "undesirable political aspect attendant" upon litigation could be avoided, the suits, he said encouragingly, could be "filed and rushed to a speedy conclusion."[25]

The Justice Department concurred with Johnson's recommendations, and the next several months were used for a "thorough preparation." The Bureau of Forestry and Geology proceeded with surveys of the affected area. The Bureau of Chemistry carried out more investigations of the forest. Studies were conducted on methods of removing sulphur from smelter smoke and the "commercial practicability" of converting it into fertilizer. Johnson, as counsel for Georgia, also determined to "direct the smelting situation" at Ducktown and Copperhill in order to provide "the fullest example of the feasibility of extensive operations without damage."[26]

J. K. Haywood of the Bureau of Chemistry and W. J. Hughes of the Department of Justice toured the West. They visited Denver, Salt Lake, and the smelters near San Francisco in order to familiarize themselves with conditions around these smelters and with the facilities in operation at the plants before going to the Mountain Copper Company smelter near Redding. At Redding, they asked that the current damage be recorded so that further damage could be identified and distinguished in the event settlement was reached in the Mountain Copper case. Then the two men visited Butte and Anaconda. There Haywood's investigation revealed that the damage had increased since his investigations of 1906 and 1907. Shortly afterwards, Gifford Pinchot and Secretary of

Agriculture James Wilson visited the Anaconda region and inspected the Deerlodge National Forest and the Deer Lodge valley.[27]

In the meantime, Johnson had settled the Mountain Copper Company litigation, avoiding a damaging legal precedent against the federal government. Settlement was negotiated on the basis that there would be no repetition of the injury, receipt of ten thousand dollars for damages, and agreement to a stipulation obligating the company to cover any future loss incurred on about six thousand acres of national forest if the land failed to reforest. At the same time, they entered stipulations of record with the approval of the Supreme Court, which Johnson claimed "largely overcame the precedent of the antagonistic decision."[28]

Upon conclusion of the Mountain Copper Company case, Ligon Johnson began preparing a draft of the charges to be filed against the Amalgamated. The bill was written and philosophically justified within the framework of conservation principles. The most dramatic articulation of this was provided by Johnson in an exchange of letters with Attorney General Bonaparte. In discussing the nature of the suit, Johnson asserted that

> The action proposed by the bill in equity against the Amalgamated Copper Company is . . . in furtherance of one . . . , if not the most important, policy of the Administration and the Government, which policy is the conservation of the nation's resources and particularly of its water supply.
>
> The smelter is located on the very edge of the National Forests. . . . The surrounding country . . . prior to the unrestricted liberation of smelter fumes [was] fairly well forested and covered with vegetation.
>
> The smelter was placed in its present location exclusively for the convenience of its owners. . . . The choice of the site was deliberate and with full knowledge of the damage which would result from the operation of the smelter.

Johnson attributed the devastation to the irresponsible economic practices of a "commercial Hercules": "In the fifteen or twenty years of smelting at Butte all vegetation has been killed and the surrounding mountains denuded, and, naturally, the water supply in the immediate vicinity having failed and become insufficient to furnish the needs of the smelter plants, it became necessary to bring water to the smelters, or to take the smelters to the water." The Amalgamated, Johnson said, chose to take the smelter to the water because it was cheaper and simpler than making heavy investments in the "old-fashioned" Butte smelters and resolving difficult engineering problems.

> A new, combined, and up-to-date plant was much to be desired. And, since a new plant was warranted, it was much more economical for the company to build it upon a watershed having an abundant water supply, although the new location involved a haul of ore from Butte to the new site, letting the Government and the farmers suffer the damage from its operations, than to re-build in the region already converted into a desert waste when such rebuilding would require expensive dams and long and costly aqueducts.

Unless the Amalgamated was restrained, a "perambulating agent" of destruction would continue inexorably from one site to another:

> As the smelters were dismantled at Butte and moved to a place where the water supply had not been destroyed by them, so, once more smelters may be dismantled at Anaconda and a new site chosen. The perambulating agent of destruction will thus again begin its search for a new site, and green fields not yet slighted and made desolate by its presence will be rendered useless. The abandoned locations, however, once made desolate by these furnaces must always remain so . . . and a monument of desolation to the end of time left standing to the economical policies of a commercial Hercules, which claims immunity from responsibility, and the right to non-interference upon the very ground of its strength and commercial importance.[29]

Drawing upon their experience in Tennessee and California, the administration concluded that only the federal government as sovereign could force the Amalgamated to control its emissions, because a "complete remedy will require an outlay by the copper company of so large a sum that only a decree of court will bring about such an expenditure." The investigations of Johnson and of the various subdivisions of the Department of Agriculture had convinced them of the deadliness of smelter emissions and the proposition that the poisons could be eliminated. "Sulphur dioxide is the most destructive chemical agent known to science, so far as vegetation is concerned," they agreed. Echoing the opinion expressed by Dr. Heber Robarts almost two decades earlier, they said, "Physicians certify to its deleterious effect upon the human system, particularly as to those persons suffering from any affection of the air passages." And they felt the effect of arsenic on the human system, although not "fully determined," was "doubtless injurious" in view of the large numbers of cattle that had died "within the limits of arsenic distribution." Further, flue dust was "harmful to vegetation, unpleasant to breathe and [had] a sterilizing influence upon the soil."[30]

Johnson's involvement in the operations at the Tennessee plants over the past year had demonstrated that the prevention of the fume injury was "no longer an open question." As a result of the Tennessee experience, not only had poisonous substances been removed from the smoke, but a new industry had been created. A "new and cheap supply of sulphuric acid" had been furnished the fertilizer industry. The operation of the Tennessee plants had been inspected by the Bureau of Chemistry, and it had concluded that "there is no longer any question of the scientific or commercial feasibility of the manufacture of sulphuric acid from waste fumes." The optimism of Roosevelt's men on this point was strengthened by a report in the *Intermountain Republic* headlined "Tennessee Copper Co. Turns Smoke into Profit." The report explained that a market for fertilizer had been found among farmers in the South, and since then Tennessee Copper stock had increased 10 percent. All that was required now, they thought, was to implement the suit and compel the Washoe smelter to institute abatement methods that would not only work to the benefit of the land and resources, both human and material, but also to the economic interest of the Amalgamated.[31]

By the middle of 1908 a draft of a "bill in equity" was completed. The proposed suit was drafted as a bill in equity on the basis that the federal government had a right "in its sovereign capacity to proceed towards securing an abatement of the sulphur fume injuries" in the Anaconda region. "The fumes," the draft read, "being a public nuisance upon a National Forest are subject to abatement at the insistence of the Government."[32] A suit in equity was chosen because it was more pliable in court and the judgment or decree could be molded into a form that suited the exigencies of the case. The suit was characterized by Solicitor General Henry M. Hoyt and Attorney General Bonaparte as "novel and important," and its various ramifications were the subject of discussions and analyses during the entire time of its preparation. One of its most novel elements was the question of the proper assessment of damages. W. G. Weigle, assistant chief of forest management, who had carried out an extensive investigation in the Deerlodge National Forest, was asked for his recommendations on the problem. Weigle felt damages of three types should be assessed: (1) the cost of reforestation, (2) destruction of the quality of the soil, and (3) loss of "rental value." By rental value he meant "The amount of *Annual Revenue* that each acre would produce under nominal conditions."[33] Writing from Anaconda, Weigle elaborated further on the concept:

> What I mean by this is, that every acre of forest land is capable of producing a certain amount of wood annually, and every acre of grass land helps to support a certain number of sheep or cattle, for

> which a certain fee is charged. The annual wood accretion of forest land of medium quality is not less than 1/2 cord per acre, which at current prices, is worth 50 cents. Therefore, if an acre of 20,000 acres of forest land is incapacitated from making this amount of growth, the loss sustained by the Government is $10,000 annually; if this loss be reckoned in saw timber, which is a higher class material than cordwood, the loss sustained will be considerably greater.
>
> The same is true along the line of grazing. Whether the range grass within the limits of the National Forest contains sufficient arsenic to make it dangerous to stock, remains to be seen through chemical analysis, but the truth is that the stockmen are afraid of it and in consequence large areas of fine looking range west of Anaconda remain unused, thus causing an annual loss to the Government of several thousand dollars.

In qualifying his idea of the destruction of the soil's quality, Weigle explained,

> In addition to the rental for the actual productive capacity of the soil there should be added a charge for the constant diminution of the elements which make up the quality of the soil. The damage along this line is caused by the death of both large and small timber, which permits the sun's rays to burn up the humus, which is the principal support of tree life. After the humus or soil covering is gone, erosion sets in on the hillsides, and the sub-soil or bedrock is what remains.

As to the cost of reforestation, which he thought was "right and proper, but possibly new to the Courts of this country," Weigle reasoned that damages should be awarded "equivalent to the cost of planting":

> There are several thousand acres within the Deer Lodge National Forest where not only the large trees, but all of the small reproduction has been killed by the smelter fumes. Even though the smelter would cease being the cause of damage to vegetation today, there being no natural reproduction left on the ground and no seed trees, within a reasonable distance, there is no possible way

> to reforest these areas except by planting, which will cost the Government at least $8.00 per acre. The smelter is the cause of this condition, therefore why should they not be called upon to pay the cost of reforesting?

Weigle estimated that only a part of the timber had been killed by sulphur dioxide "up to the present time," but he asserted he had "seen sufficient to know that the stumpage value of the timber will amount to more than fifty thousand dollars."[34]

Justice Department lawyers spent a good deal of time refining Johnson's draft. W. J. Hughes, the attorney in charge, sent a copy of the draft to R. L. Clinton, counsel for the Deer Lodge Valley Farmers' Association, requesting that he read the bill carefully and give them the benefit of his criticism as a result of his experience with the Amalgamated attorneys in the Bliss case. Hughes had discussed the case personally with Clinton in Butte, and Clinton had told him then that they had better prepare for an unusually rugged fight inside and outside the courtroom. For example, Clinton told Hughes, as a consequence of his involvement in the Bliss case he could not "even trust his stenographer in regard to writing about the litigation."[35]

Clinton found the bill "very complete" but amplified certain parts, explaining, "I find every advantage in having framed comprehensive pleadings as the defendants are very technical and will object very strenuously to the introduction of any evidence which would not be allowed if a close technical ruling was made in their favor; that is to say, they will insist with very great energy that no proof go in unless licensed by exact and precise language of the Bill."[36]

In August 1908 Solicitor General Hoyt decided, in spite of their perfected bill and impressive evidence at that time, to strengthen their case still further. He wanted to corroborate the views of Weigle and others with independent investigations by the most competent men the forestry bureau could find. He considered it a matter of the "highest importance" and urged that the investigations be made as soon as possible. Shortly afterwards, the Justice Department received a communication from the White House that the president wanted to see the bill and its supporting data. Before taking any "actual steps" in instituting the suit, Theodore Roosevelt wanted to give the matter "his personal attention."[37]

– Chapter Nine –

COMPELLING A CORPORATION TO DO ITS DUTY

W. J. HUGHES BROUGHT THE AMALGAMATED FILE to the White House on October 21, 1908. Senator Joseph M. Dixon was with the president in his office when Hughes arrived. The senator had been told of the impending suit by a "thoroughly alarmed" businessman in Anaconda and had come to protest against institution of the suit. Dixon thought it would be a "most colossal mistake" because the town of Anaconda relied solely on the smelter, and the interests of the state were largely dependent upon it. The ore had to be smelted somewhere, he argued, and it could not be done without the discharge of fumes. The Amalgamated had spent $2 million in erecting a plant that "would reduce the injury to a minimum." The conversation concluded with the president's statement that if he decided "to do nothing in the matter" Senator Dixon would not hear from him. If, however, he contemplated action, the senator would have an "opportunity to be heard by the [Justice] Department."[1]

Roosevelt was little concerned about political repercussions in Montana. It was his last term, and he had little reason to curry popular favor. More importantly, he and other Republican strategists had concluded that, in any event, they would lose Montana and other western states; those losses, they reasoned, would be more than compensated for by gains in the Midwest and East. Within weeks, Roosevelt had concluded that action had to be taken in the Anaconda case, and Senator Dixon was indeed given the opportunity to be heard by the Justice Department. On December 3, 1908, Dixon met with Roosevelt and Attorney General Bonaparte. Bonaparte outlined the problem, how gravely they viewed it, the nature of the suit, and their proposed remedies.[2]

Roosevelt had decided that before initiating suit he would endeavor to mediate the matter in the hope of avoiding a burdensome and malefic legal battle. He had asked the representatives of the Amalgamated and the farmers and the Montana delegation to meet with the leaders of his administration. Roosevelt, as leader of the Republican Party and the nation, had developed an exceptional range of devices that he had used with recalcitrant opponents of his administration's policies. One of the most perceptive students of the period described these devices: "Special conferences, dramatic investigations, public condemnations, and private encouragement all found a place in the master's repertory," he said. To contend with the Amalgamated, Roosevelt chose the "special conference" technique in the hope of maneuvering the parties into an accommodating attitude. They would all sit down like gentlemen and discuss the affair. In the fashion amenable to nineteenth-century politics, bargains would be struck, agreements reached, and the problem resolved.[3]

On Saturday morning, December 5, 1908, Roosevelt presided over the meeting at the White House. In attendance were John D. Ryan, president, and A. J. Shores for the Amalgamated; Nick Bielenberg, W. C. Staton, and attorney R. L. Clinton for the Deer Lodge Valley Farmers' Association; Attorney General Charles J. Bonaparte, Solicitor General Henry M. Hoyt, Special Assistant Ligon Johnson of the Department of Justice, and Chief Forester Gifford Pinchot for the administration; and Montana Senators Joseph M. Dixon and Thomas H. Carter on behalf of their constituents. Senator Henry Cabot Lodge also attended, reported the Butte *Inter Mountain*, "voluntarily in behalf of the thousands and thousands of stockholders of the company who reside in Massachusetts."[4]

The discussion focused on remedies the Amalgamated could undertake to prevent further damage to the forests and public domain. Roosevelt, after studying the reports of the forest service and the Haywood investigations, had concluded that, beyond question, damage had been established. What the president wanted was agreement on a remedy. The administration felt there were three alternatives: (1) move the Washoe smelter to Great Falls and merge it with the Amalgamated smelter there, (2) move the incinerating operation of the smelter back to Butte, or (3) convert the Washoe's sulphur dioxide fumes into sulphuric acid and manufacture fertilizer.

In proposing the first alternative, the administration argued that the water supply in Great Falls was sufficient, that the only vegetation affected would be grass, that there was no timber within a considerable distance, and that the region was fairly level and consequently erosion would not be as severe as it was in the mountainous region of Butte and Anaconda. Therefore, the water supply would not be as severely damaged, the forests saved, and the damage to crops in the Deer Lodge valley ended.

The second alternative, moving the incineration operation back to Butte, appeared to be the simplest remedy in the administration's eyes. "Practically all the damage" had been done that could be done in the Butte region, Ligon Johnson had concluded in his report. And the investigation had revealed that a considerable portion of the operations of the Anaconda plant was devoted to the concentrations of ores for the incinerating process. Three-fifths of the ore delivered to the smelter was concentrated in the preliminary operation and only two-fifths was left to be incinerated. That two-fifths, the administration argued, could be shipped back to Butte over the same railroad and incinerated there. As a result, widespread fume damage to the forests and vegetation would be prevented, erosion curtailed, and stream pollution and crop damage reduced.

In pressing their point, the administration confronted John D. Ryan with the image of his old nemesis, F. Augustus Heinze. Johnson pointed out that before the consolidation of the Butte smelters and their relocation to Anaconda, Heinze had treated his ore by hauling it from Butte to Basin, a distance of some twenty miles. After concentrating it in Basin, he had hauled it back over the mountains to Butte for the incinerating process. Heinze's haul was "infinitely more difficult" because it had to contend with the severe grade of the Continental Divide. The report before the president on this proposition concluded that "The only additional cost entailed upon the company being the return haul from Anaconda to Butte, and the removal of the roasting furnaces. This return haul to Butte . . . would be a much shorter haul and a much less expensive haul than that necessary in handling the Heinze ores prior to the dismantling of the Butte smelters and the erection of the Anaconda plant."[5]

The third alternative, conversion of sulphur dioxide into sulphuric acid for the manufacture of fertilizer, called for implementing the method being employed successfully at the Tennessee smelters. That which was not used for fertilizer, administration spokesmen held, could be merchandised to the manufacturers of high explosives and smokeless powder, of steel and acids and paper, and to all the other enterprises that used sulphur dioxide in the preparation or refinement of foodstuffs and oils.

It was this alternative, the production of fertilizer, that the administration felt was the most compelling proposition. Justice Department officials pointed to the increasing consumption of fertilizer in the South. It was "doubling about every six years," they reported. They recognized that little fertilizer was currently being used in the West, but attributed that to freight rates. If the southern states were located in relation to the sources of fertilizer as were the western states, the Department of Agriculture had determined, they too would probably use no fertilizer, "the cost being too high to bring adequate returns for the outlay." But since the U.S. Geological Survey of 1907 showed that "the largest and richest phosphate deposit in the world" was located in Idaho, "close to the Washoe smelters," freight rates would no longer be a problem for the smelters.

Fertilizer is produced by mixing sulphuric acid with phosphate rock. The Justice Department's report to Roosevelt pointed out that this "undeveloped and practically useless asset, phosphate rock, and a harmful fume, . . . if utilized together [could] eliminate the injury complained of and mean much in the development of the western country and particularly of its agriculture." In addition, the Division of Soils had assured the department that if fertilizer could "be had at a cheap enough figure it would be generally beneficial to the western soils and highly profitable to the farmers." It was the Washoe's good fortune, administration officials claimed, to be so auspiciously situated that they had the opportunity of availing themselves of a promising commercial opportunity while simultaneously abating the fume problem.[6]

Amalgamated officials were unimpressed. Moving the incinerating operation to another location was out of the question in view of the $10 million expended on the Washoe smelter to date. The methods employed at Tennessee to remove sulphuric acid were not feasible for the Washoe plant, they said. They pointed to the testimony of their expert, Frederick J. Falding, in the Bliss case that the "Ducktown process" was "impractical" for the Washoe smelter. Falding had claimed the "problems at Anaconda were not at all like those at Ducktown," the Washoe being "at least six times as large as the plant at Ducktown." The business of manufacturing fertilizer, said the Amalgamated officers, was impossible because there was no demand for fertilizer in the West, and furthermore, the phosphate beds, which Mr. Johnson discussed, were located at Montpelier, Idaho, a distance of 380 miles from the Washoe. "At [current] freight rates," John D. Ryan claimed, "the cost of phosphate rock handled this distance would be prohibitive." Once again, company officials complained about the severe economic hardship an injunction would work on the state and workers of Montana. In addition, they claimed, "practically the whole copper producing industry would be brought to a stop."[7]

Roosevelt was not unimpressed with the arguments of President Ryan and his spokesmen. After the meeting, he told Bonaparte that the allegations "made by seemingly responsible people . . . should as a matter of course receive our careful consideration."[8] Roosevelt and his attorney general had a number of conferences following the meeting in which they discussed the proposed suit. They did not regard it without trepidation. There were a number of inherent problems that augured ill for prosecution of a suit that looked primarily to forcing a remedy rather than a punishment. Four problems stood out:

1. First of all, Bonaparte explained to the president, the conversion of sulphur fumes to sulphuric acid and then to fertilizer was a new process, and marketing required "the application of scientific principles to actual business conditions," a practice just beginning in this country.

2. There was a complex of trusts within the industry. To assess properly the "practicability" of entering the fertilizer field they would have to have "expert business reports" as to market conditions and plant costs. This information, Bonaparte said, "will probably be hard to obtain, because it is all in the possession of the fertilizer trust, which is of course, hostile to the government." At the time, the outstanding representative of the fertilizer trust was the American Agricultural Chemical Company. It had been organized in 1893 and had numerous plants and other properties in the chemical world. It had "intimate connections" with the "house of Morgan," in which "a considerable part of the credit facilities of the

country, were being concentrated." Conspicuous in the inner group of financiers, of which the house of Morgan was the leader, was the National City Bank of New York. That bank, along with Standard Oil, of course, had figured prominently in the formation of Amalgamated. Within the next decade the American Agricultural Chemical Company and a smaller organization, the Virginia-Carolina Chemical Company, would control about one-third of the total supply of fertilizer and chemicals in the country. Seven other corporations would account for the remainder. Through interlocking directorates, which established interconnections among and between large industrial and financial concerns, common interests and general objectives could be—and were—pursued.[9]

3. Another problem closely allied with the complex of trusts was the belief in their omnipotent power to determine the course of American life. "There is some reason to think," Bonaparte observed to Roosevelt, underestimating the situation, "that the practical experts who know most about the matter will be indisposed to help the inquiry because business interests allied with the Amalgamated or hostile to the Government control the situation everywhere both as to the smelting of copper and the manufacture of fertilizers." The Justice Department was under no delusions. "The mouths of all the experts would be closed," Bonaparte explained, "because the Amalgamated and the Standard Oil Company ultimately will reach them and control them wherever throughout the country and in whatever individual enterprises they are engaged; that is to say, practically they are already in the employ of the Standard Oil and its allied interests."[10]

4. Finally, the administration was apprehensive about the "Montana situation" itself: "The suit will be filed," Solicitor General Hoyt said, "in a territory which has heretofore been the battleground between the Amalgamated and Heinze interests," and "where open confessions of sales of political and even judicial influence were lightly looked upon."[11]

Because of these problems and because of their wish to create a cooperative partner rather than a hostile adversary, Roosevelt and Bonaparte agreed to process the suit in a manner that would allow the Amalgamated the greatest latitude in avoiding economic hardship and still allow the company time to find the method that would abate the fumes. "Without doubt," Bonaparte told Roosevelt, "the prayer for an injunction could be in the alternative." They could ask for "either an injunction or the construction of a plant to consume the fumes and remove the evil." They could file a stipulation in the case wherein the company "would agree to build such a plant within a reasonable time, to be agreed upon." This "would be as satisfactory, I suppose, and as effective," Bonaparte concluded reflectively, "as to proceed with the bill *in invitum*."[12]

In a letter to the attorney general on December 9, 1908, President Roosevelt stamped his personal style on the "Anaconda smoke case" and expressed his deep concern about the effect of the case on the future of the movement to control smelter emissions. "It seems to me imperative that there shall be a full and careful investigation" with respect to this suit, he said. "To follow any other course would be, it seems to me, to act without regard to the interests of the people of Montana, and elsewhere." Investigations were in process in California. Roosevelt wanted to be confident that the anticipated benefits outweighed the costs: "My directions in connection with the Anaconda smelter matter are that we shall look carefully before we leap," he instructed.

> No suit of this size and importance should be entered into by the Government until it has fully and carefully thought out what the consequences will be. . . . The Government is bringing the suit on the ground that the forests and grazing lands, and incidentally the waters, in the neighborhood of the smelters are destroyed. Before bringing the suit we should as mere matter of common sense find

> out whether if successful we save some destruction at the cost of destroying twenty times as much property ourselves.

The case was pivotal if they were to proceed against smelters on a broad front. If they could not win in Montana, their cause was in serious doubt elsewhere.

> The suit is especially important because of its relations to many other suits of the same kind that can be or should be carried on in other parts of the country. I wish to know the facts before proceeding. I regard the fact of the damage to vegetation within a certain area of the smelter as established. I do not regard as established the allegation by the defendants that the smelters would have to stop completely if the suit was successful.[13]

Having taken note of Johnson's counsel that threats would be made to close down and "Promises will be made to be broken, if [the Tennessee] experience is to be believed," Roosevelt declared, "I pay heed to the statement made on behalf of the Government that unless these parties are forced to by law, they will not stop their destructive proceedings, because expense will be involved in stopping them." While appreciating the coercive factors required, he was, nevertheless, determined not to harass the industry: "I wish to know something about the expense, whether it really does represent prohibitive expense, which would mean the closing up of the works, or whether it merely represents a heavy necessary expense which can and should be borne and which will allow the work to be done, but only under conditions that prevent its being noxious to the vegetation round about."[14]

Immediately, Bonaparte initiated the "full and careful investigation" requested by the president. In a meeting with Johnson and Hughes, he told them what Roosevelt wanted: "What the President wishes to know is the practicability as a matter of business and not as a theoretical or scientific matter of conversion of fumes into sulphuric acid for the treatment of sulphate acid for the treatment of sulphate rock, and the cost of installing such a plant, both in general and at this particular place."[15]

In order to determine the "practicability as a matter of business," Bonaparte outlined what the investigation should entail. He wanted information on the methods of making fertilizer and on the use of fertilizer in the West. He wanted information on the "cost and practicability" of having the Washoe install a smelting plant modeled after those in Tennessee. He wanted more information on the idea of moving the smelter "back to Butte or to a place where there are no farms or forests to be injured." The investigation should also include "something," Bonaparte said, on "the capital and resources of the Amalgamated, in order to show that even an expenditure of seven or eight million dollars would not be a serious burden upon them."[16]

At the same time the Department of Agriculture initiated a comparative investigation of the national forests in Montana and Idaho under the direction of Dr. George Grant Hedgcock, plant pathologist of the Bureau of Plant Industry.[17] And Gifford Pinchot, the chief forester, engaged the services of Profs. George F. Peirce and R. E. Swain of Stanford University. Swain was associate professor of Physiological Chemistry and Peirce was professor of General Botany. Swain had assisted Prof. W. D. Harkins in the investigations of the Deer Lodge valley on behalf of the farmers in the Bliss case. Peirce and Swain were both members of the San Mateo (California) Home Protective Association, which had been organized to oppose the erection of a large smelter on San Bruno Point, land owned by the Guggenheim trust. The Home Protective Association was composed of "many rich and prominent people" of San Francisco, according to a local newspaper, and the smelter site was situated several miles from Stanford University at the present site of the San Francisco International Airport.

Professors Peirce and Swain had visited the large smelters throughout the country from Ducktown, Tennessee, to the Pacific coast, conducting field and laboratory tests to determine the effects of smelter fumes on vegetation. The San Mateo Home Protective Association had based its opposition to the smelter on the data of the two professors and warned the Guggenheims that the operation of the smelter would be "bitterly fought in the courts." The Guggenheims had expended, according to the newspaper, over a million and a half dollars before finally seeing "fit not to antagonize" the property owners of the San Bruno area. The professors, with the authorization of the association, turned over the results of their studies and experiments to the Department of Justice.[18]

As the officials of the Roosevelt administration energetically set out to prove the practicability of their remedies, Amalgamated officials likewise set out to prove their impracticability. Attorney C. F. Kelley, now the general counsel of the Amalgamated Copper Company, submitted two chemical reports prepared by their experts, Charles A. Doremus, "eminent toxicologist and laboratory chemist," and Frederick J. Falding, a "most capable engineer." Investigations by Ligon Johnson revealed that Doremus sought "to compare conditions at Anaconda and Tennessee, when he [had] never been to Tennessee," and Falding, who had "never seen the Anaconda plant," sought to compare it with the Tennessee plant. He based his report "exclusively on figures furnished and gas estimates supplied by the Amalgamated Company. . . . Nothing," Johnson complained, "is given of his own knowledge."[19]

To further complicate the issue, Johnson had before him "four separate sets of figures," each one different from the other, yet all dealing with the same question. One set of figures was from Doremus and another from Falding. Dr. Doremus's figures were "radically different from the figures of Mr. Falding," and figures from general counsel Kelley were "different from those of both Dr. Doremus and Mr. Falding." The fourth set was that of the sworn testimony of the Amalgamated officials in the Bliss case, which in turn differed from those of Doremus, Falding, and

Kelley. "Which of these four sets, if any," Johnson remarked to Bonaparte, "are correct I am, naturally, unable to state." These so-called reports were "merely argumentative statements for the Amalgamated" and as such were "not entitled to great weight in the matter of determining the Anaconda situation," Johnson maintained. Happily, however, upon examination and analysis of Falding's and Doremus's briefs, he said, they "actually bear out the contentions of the Government that acid may be made and the situation remedied by the installation of acid plants."[20]

As a result of his intimate knowledge of the Georgia-Tennessee copper company suits and of the method of converting sulphur oxides into sulphuric acid for fertilizer employed there, plus the government's investigation to date, Ligon Johnson was able to pick apart Doremus's and Falding's reports, revealing in the process numerous "inaccuracies and misrepresentations." In Doremus's report there were many inaccuracies but there was one sufficient to discredit the entire statement: "Neither fumes nor arsenic," Doremus stated, "did damage around Anaconda." Delighted to find this falsehood, Johnson pointed out to Bonaparte that "Since the hearing before the President we have received voluntary treatises, accompanied by photographic evidence, from Prof. George J. Pierce [sic] . . . and statements by Prof. Harkins and others, verifying under separate investigations" the declarations by the various government experts.[21]

Falding's report controverted his sworn testimony in the Georgia-Tennessee copper cases. He claimed that a reduction of sulphur dioxide at Anaconda would not "materially change the conditions as to the effect of the smelter fumes on the surrounding country." How, Johnson asked, was he to reconcile this statement with his assurances on behalf of the copper companies in Ducktown, Tennessee, to the officials of the State of Georgia that removal of comparatively small amounts of sulphur from the smelter's emissions "would remedy the evil in Tennessee"?[22]

As to how the Amalgamated experts' reports supported the government's contention that sulphuric acid could be made "practicably" at the Washoe, Johnson concluded that it was not necessary to eliminate *all* the sulphur dioxide in the smelter fumes: "It is generally conceded," he claimed, "that the condensation of approximately 65% of the gas would afford relief at plants similar to those at Anaconda and Tennessee." The total sulphur given off daily in the form of gas, as reported by Doremus, he said, was

McDougal roasters	432 tons
Blast furnaces	202 tons
Reverberatory	30 tons
Converters	138 tons

"In other words," Johnson told the attorney general, "considerably more than half the gas is derived from one set of roasters—the McDougals." Dr. Doremus, he continued, explained in his report that acid could be made from the McDougal furnaces, and furthermore Doremus had asserted that the ore at Anaconda was higher in sulphur content and weakened less by the influence of other gases, such as coke gas, than were the gases of the Tennessee copper companies. Falding's report made the same admission, Johnson concluded exuberantly. In the Georgia-Tennessee case it was established that McDougal furnaces could be adapted to the production of acid. Therefore, by merely modifying the McDougals at the Washoe plant, more than 50 percent of the sulphur dioxide could be converted into acid and removed from the emissions, and certainly some means could be found to reduce the sulphur emissions to some extent from the remaining furnaces, the reverberatory, and the converters. In view of the inconsistencies in the experts' reports and the promise that they would be pinned down and a commitment obtained, Johnson and Hughes were sent to New York to meet with general counsel Kelley and his experts.[23]

In New York, Johnson and Hughes were able to obtain a number of admissions on the capability of the Washoe plant to produce acid. Falding "frankly admitted," said Johnson, that McDougal roasters had been adapted to the production of acid and that the gas from them at Anaconda could be converted to acid. Falding further admitted that it was "more than probable" that the 138 tons of sulphur dioxide emitted from the converters could also be condensed into acid. In addition, Johnson reported to Bonaparte, Falding stated that "he could not say that an altered method of smelting at the blast furnaces would not permit the production of acid from those furnaces."[24]

Dr. Doremus "made practically the same admissions." Johnson had called Doremus's attention to the fact that in his report he had asserted "that acid could be made from a fume of the strength of 1.3%" and that the Washoe company itself testified that the strength of the fume as far away from the furnaces as the big stack was 1.4 percent. At this point Johnson thought he almost had the problem solved. Enthusiastically, he reported to Bonaparte, "The very experts offered to establish the contrary, affirmatively show that the acid can be made as asserted by the Government. In other words, acid could unquestionably be made from the 423 tons of sulphur daily given off from the McDougals, could probably be made from the 138 tons eliminated through the converters, and the 232 tons remaining might possibly be condensed under modified operations."[25]

General counsel Kelley wanted to "file for consideration" the preliminary findings of the court in the Bliss case. Hughes and Johnson, however, told him that these findings were "in no wise pertinent." That suit they countered, "applied only to the protection of low lying valley farms, and damage to cattle and forage by arsenic was the chief matter discussed."[26]

Then Doremus focused on another matter—"marketability." "The chief question was not whether or not the acid could be made but what was to be done with the acid when it was made," he argued. Transportation of acid would require a "large outlay for tank cars which had to be returned empty," and because of the danger and difficulty in handling acid, "rates to Chicago and the East would be prohibitive." Johnson replied that since a large deposit of phosphate was situated near Anaconda at Montpelier, Idaho, the acid could be converted into superphosphate at the Washoe and then "handled on much the same basis as ore." On the question of "marketability" that discussion ended.[27]

In a series of conferences that followed, involving Attorney General Bonaparte, Solicitor General Hoyt, Johnson, and Hughes, Amalgamated representatives endeavored to convince the Roosevelt men that "only the fumes from the McDougal Plant alone should be considered in connection with making acid" because it "was impossible, or at least impracticable to make sulphuric acid" from the other parts of the smelting plant. Narrowing the "matter of desulphurizing the fumes," they argued, involved three fundamental questions: First, whether "as a practicable commercial proposition" sulphuric acid could be made from these fumes. Second, if it was possible, what was the cost of constructing and implementing the process? And third, what disposition "might be made of the product?"[28]

Writing Solicitor General Hoyt in late January 1909, general counsel Kelley stated in broad and starched terms just how far the Amalgamated was willing to go in removing poisonous fumes at the Washoe:

> I stated to you that I thought my Companies would agree that if upon receiving competent engineering and chemical advice, after full investigation and consideration, they were assured that as a commercial proposition it was possible to make acid, to obtain title to phosphate beds of sufficient area and extent to justify the enterprise, and further, if they could be assured upon investigation that there was any market for the disposal of the product, that they would not hesitate to engage in the enterprise.
>
> After consultation with Mr. Ryan and Mr. Thayer I beg to say that this position has been confirmed, and that if upon investigation it

> can be determined that the enterprise can be commercially carried out we will readily agree to do what we can toward removing any controversy which may exist between the Government and our Companies, by building such plant.
>
> The foregoing involves considerable investigation, and we have submitted the matter to Mr. Falding who will advise us as to the proper course to pursue, and who will also cooperate with any person designated by the Government to assist us in solving the problems presented.[29]

Attorney General Bonaparte, recognizing an opportunity to gain an advantage, however small, wrote Kelley almost immediately. He wanted to clarify the understanding regarding the abatement of fumes emitted by the smelter as a whole and, more importantly, to reach a "final agreement" on modifying the McDougals and defining the limits of the investigation for which the government would be responsible in determining the "practicability" of converting the sulphur dioxide into sulphuric acid. "Sir," Bonaparte wrote crisply,

> The Department is in receipt of your letter of January 23, addressed to the Solicitor General. It is the understanding of this Department that you stated:
>
> (1) That if any competent chemical engineer, after due investigation, shall give a professional opinion establishing the practicability, not only scientifically but commercially, of converting the sulphur fumes from the McDougal plant at the Anaconda smelters into sulphuric acid, you will accept this decision. . . .
>
> (2) The Department understands that any question of extent of injury to public interest from the continuance of the fumes on the one hand, or of loss from interference with your private enterprise on the other, are now immaterial, at least for the purposes of this correspondence, and that the question whether markets can be found for fertilizers produced from such an acid manufacturing plant is also irrelevant for present purposes, so that the only points

> requiring any investigation or as to which you ask assurance are the extent and quality of the phosphate beds in Montana, not very far from Anaconda, to which Mr. Johnson called your attention from the reports of the Geological Survey, and the possibility of your obtaining a satisfactory title to the same.[30]

In regard to the administration's understanding of the capability of the smelter to abate all the fumes at the smelter, Bonaparte continued,

> Your expert, Dr. Doremus, and those consulted by the Government agreed that a removal of approximately sixty percent of fume strength would abate the injury. The figures compiled by Dr. Doremus showed that this elimination would be accomplished by a condensation of the McDougal fumes; for this reason, and not at all because the fumes from other appliances were thought incapable of conversion, only the McDougals have been considered.[31]

Anticipating the use of Frederick Falding as a means of playing for time and frustrating the administration's efforts, Bonaparte designed his words and punctuation carefully: "The situation is very materially changed if you now deem it necessary to refer the entire matter to Mr. Falding for his advice as to your proper course for you to pursue, and intend only that he, Mr. Falding, shall cooperate with any person designated by the Government 'to assist [the company] in solving the problems presented.'" In concluding, Bonaparte stated:

> Having thus explained the Department's understanding of the results of the conference referred to, I need only add that the Department will lay before the President the results of the investigation so far made, attaching to its reports your letter [to] which this is a reply and the reports to you from Mr. Falding and Professor Doremus, and that Mr. Johnson, special counsel of the Government, will soon proceed to Anaconda to pursue the investigation.
>
> Further, the Government will investigate, as soon as the season permits, the location and availability of the phosphate beds in question, in which inquiry it is understood you will cooperate.[32]

On the same day he wrote Kelley, Bonaparte dispatched a long report to President Roosevelt, bringing him up to date on the status of the case by summarizing all the pertinent reports and his recommendations as to future action. "It seems more than likely," Bonaparte observed, drawing attention to a condition that would loom ever larger in neutralizing the government's efforts, "that the Tennessee Copper Company and the Amalgamated Company are allied, embrace the same industrial and financial interests, and what can be done at one plant with commercial success can be done at the other." Confidently, he expressed the opinion that he did not "doubt that if the manufacture of sulphur fumes into sulphuric acid for further manufacture into fertilizers has been successfully and profitably inaugurated as a going commercial enterprise at the plant of the Tennessee Copper Company" at an installation cost of $800,000 or less, it could profitably be done at Anaconda. Revealing what "commercial practicability" meant to him, he observed that "Five industrial plant units of the same capacity and costing proportionately less will dispose of the fumes from the McDougal plant . . . and will do so on a profitable basis and at a cost of construction of less than four million dollars."

Phosphate beds "of large extent and very good quality," he said, "are situated so near the operations of the Amalgamated Company at Anaconda that there can be no question of availability." The only point to be determined more accurately by government exploration was the extent and quality of these beds. He had "overwhelming testimony" that transportation of the phosphate rock was no problem. Presently it was being shipped from abroad and then transported "in this country a thousand miles or more to fertilizer manufactories" and then the product sent great distances to market.

There was, Bonaparte continued, one big question the government had yet to contend with: the market, or, more accurately, the doubt in

Bonaparte's mind that the Amalgamated could be moved to develop and take advantage of a market for fertilizer. Bonaparte's investigation had convinced him that the agriculture of the West would provide a viable market within close proximity of the Anaconda smelter. "Putting it broadly," he expressed his confidence in the power of the free marketplace to provide the initiative for remedy:

> I cannot doubt that within the range of transportation from Anaconda to the Pacific coast and to the middle West as well as to the arid and semi-arid regions thereabouts, that is, within the range, say of a thousand miles within which freight rates would not burden the traffic too much, an abundant and growing use of fertilizers would be built up. Markets, if they do not exist now, would result, as is often the case, simply by the creation of a supply of the useful article.[33]

Beyond this question of a market, there was, Bonaparte admitted, "one point of immediate practical difficulty":

> The number of equipped and competent chemical engineers in this country is not great, and in one way or another most of them seem to be retained by the copper producing interests. It is necessary . . . I think, that the Government should be definitely advised by some such expert that what has been done in Tennessee can undoubtedly be done, as we believe to be the case, at Anaconda, and that the McDougal plant can be equipped, and successfully and properly equipped, with fume-consuming and acid-manufacturing apparatus.[34]

Johnson, Bonaparte advised the president, would proceed at once with further investigation to determine the potential market for fertilizer in Montana and throughout the West and would determine the status of the title to the phosphate beds and their size and quality. Meanwhile, the Justice Department would try to "find and retain a qualified expert" to work with Johnson in determining the workability of converting sulphur to fertilizer at the Washoe plant.

Turning to the proposed suit, Bonaparte avowed that the investigation was strong enough to warrant filing the suit. "In my judgment," he declared, "even if the results of the investigation . . . cannot be conclusively and finally laid before you within the next month," it was "manifestly advisable" that suit be "filed before your term of office expires." The reason he wanted the case brought to trial was because by "merely filing the suit" it would bring all the "controversial questions of fact" and "proposed remedies" into the realm of judicial consideration. There the facts, merits, and demerits of the case could be "regularly determined" in the crucible of adversary proceedings. "In other words," he emphasized to the duty-conscious but pragmatic president, "The real effect of the Government suit would not be to stop the Anaconda works, but to compel a corporation whose record as to its regard for the public interests and for its own civic duties and responsibilities is very far from clear and satisfactory, to do its duty at least in this matter."[35]

Within days, Roosevelt replied to Bonaparte's report. "It seems to me," he observed, "that ultimately we shall have to proceed with the suit to compel the Anaconda smelter people to do their duty." Nevertheless, he wanted to reserve final judgment until he received more information from Johnson and a reply from Kelley to Bonaparte's letter seeking a final agreement.[36]

In the meantime, the attorney general had already received a reply from Kelley. The letter confirmed what Johnson had suspected and had warned the head of the Justice Department about earlier: The Amalgamated would stall until the next administration in hopes of receiving more favorable treatment.[37] The officials would wait until the storm of the Roosevelt administration had passed. The administration would be there only so long, and when it went so too would the dedication and fervor of "the Roosevelt men." The trust's interests, on the other hand, would endure. The reply infuriated Solicitor General Hoyt. "This letter from Mr. Kelley," he wrote the attorney general,

> shows how little reliance can be placed upon any efforts at amicable negotiations with the representative of the Anaconda people. It shows that notwithstanding what Mr. Kelley said in the presence of Mr. Johnson, Mr. Hughes and myself, the attitude of his company will be to controvert every point and to delay as long as possible the settlement of the only question before the President, namely, whether the suit shall be brought.
>
> So far as I can see, we have not made a single step in advance in two months, and of course that suits the Anaconda company and the Amalgamated company perfectly well.[38]

Kelley would not stand on the two points requested by Bonaparte: whether the McDougals could be equipped with fume-consuming devices and whether the phosphate beds were available and rich enough to work. The Amalgamated wanted guarantees on the commercial feasibility of engaging in the fertilizer business. Kelley's letter served notice on the government that defining what Roosevelt referred to as "practicability as a matter of business" would remain firmly in the grip of Amalgamated.[39]

The next day Hoyt had not yet cooled off, and in a communication to Bonaparte he reflected the administration's numbing frustration over the inescapable reality of the power of the trust in American life:

> The attitude of Mr. Kelley and his clients simply is to my mind that the Government shall not have the temerity even to bring suit to get rid of an admitted evil, without giving a guarantee in advance that the process of removal of the evil shall be a profitable commercial venture. Indeed they go beyond this and demand in effect that the suit shall not be filed, by their refusal to enter any agreement that may be reached as to fume consumption as a stipulation of record which could be enforced, and insist that we must altogether trust them to carry out such an agreement.[40]

The next day Roosevelt, in a letter to Bonaparte, conceded to the strategy of protraction:

> In reference to the reports of the Solicitor General as to the lawsuit against the Anaconda people, I have to say that after carefully reading over the correspondence, my judgment is the same as the Solicitor General's as to the course of the Anaconda Company and the Amalgamated Company in this matter.
>
> If my administration were to continue I should direct that the suit against these companies be immediately prosecuted. But as less than a fortnight of my term remains, and inasmuch as the suit would have to be carried on under my successor, it does not seem to me wise to begin at this time.
>
> I return the memoranda submitted to me and request that they, together with this letter, be filed with the case, so that the attention of your successor on his taking office may be called to the matter.[41]

– Chapter Ten –

The Taft Men versus the Smelters

WHILE THE ROOSEVELT ADMINISTRATION HAD BEEN endeavoring to devise means to solve the smelter fumes problem in Washington, the company's press had been carrying out its established role in Montana. When word reached Anaconda that the Deer Lodge farmers had sought the aid of the president, Montanans were told of the treachery afoot by headline and editorial in both the Anaconda *Standard* and the *Daily Inter Mountain:*

"Smoke Farmers Invade Washington"

"The Shock from Washington"

"The People Are Aroused"

"John D. Ryan Has Conference with President Today"

"Mass Meeting"

"Labormen Protest against Injunction"

"Protests Pouring into Washington"[1]

Anticipating a farewell to the do-as-you-please approach to mining and smelting, the *Standard* wondered whether the "plaints of a handful of citizens" and concern for "a few square miles of forest" should be allowed to "turn back the clock of progress." "It is particularly unfortunate that less than half a dozen men could bring about such a situation at the present time," the paper editorialized. "The people have had much to contend with in the last 12 months, and as they were about to enter upon a period of seeming prosperity, along comes this report from Washington, which plunges them again into uncertainty and doubt. And to what purpose? Merely to harass and injure communities that already have suffered much? It would appear so."

Outraged, the *Standard* echoed a charge made by Amalgamated's general counsel, Cornelius F. Kelley: "It seems incredible that while court proceedings to remedy this very plaint are in progress, outside influences will be brought to bear. It seems preposterous that the livelihood of one hundred thousand people should lie at the whim of an unfriendly coterie whose claim is not substantiated by the facts nor shared by hundreds of others existing under similar circumstances." The president was going to shut down every smelter in the country, Montanans were told, unless the smelters desulphurized their fumes.[2]

With the "spontaneity" reminiscent of the petitions that flooded the state capital five years before, when Amalgamated had sought the passage of the "fair trials bill" during the war of the Copper Kings, citizens' meetings sprang up throughout the state.[3] Protests poured into Washington, said the *Inter Mountain* and the *Standard.* In Butte the protest came from the Butte Business Men's Association; in Anaconda it came from a "conference of citizens"; in Hamilton and Missoula, the cities' chambers of commerce; in Bonner and Belt, the officers of unions. The *Inter Mountain* reported that in Bozeman, "the people were ready to protest . . . although so far no call for a meeting has been made."[4]

Each of the groups that met telegraphed resolutions to Washington remarkably similar in content, opposing the "closing down of the smelters." The resolutions were recorded word for word and paragraph for paragraph in the *Standard.* At a mass meeting called by the citizens of Anaconda at the Standard hall the evening of December 5, Superintendent

A. J. Mathewson gave a lengthy explanation of "things" that was covered by the *Standard*. Mathewson threw some light on the "spontaneity" of the protest meetings. In beginning his address, he reported that "during the day he had read a telegram from John D. Ryan, which seemed to give a rather gloomy view of the situation at Washington and contained a suggestion that something be done to arouse the people interested."[5]

Whatever Amalgamated operatives did, it was sufficient to arouse the interest of representatives of both the secular and spiritual worlds. Bishop L. R. Brewer of the Episcopal Diocese of Montana, who knew, said the *Standard*, "whereof he speaks," wrote His Excellency, the President of the United States, a strong letter on behalf, he assured Roosevelt, of no one, "who has any stock in mines or smelters." Drawing attention to the inconsequential damage done to the countryside by the Washoe on the one hand and the resulting economic disaster its closing would mean on the other, Bishop Brewer pointed to an even more serious consequence: "More than that," he said, "it would kill four of my parishes and injure every parish and mission that I have in my diocese."[6]

The state land board took "official action," the *Standard* reported, and transmitted a resolution opposing the president's action and attesting that the board's investigation had revealed that the "timberlands belonging to the state of Montana are in no way injuriously affected by virtue of the operation of the said Washoe smelter." The board did *not* allude to the Department of Justice investigation that revealed that land owned by the state had been abandoned "by reason of injuries from the Washoe works" and that since the smelter operations had begun, state lands in the vicinity of the smelter had experienced a reduction in rental fees by almost one half.[7]

Although the nominally independent *Tribune* in Great Falls reported the developing story in its news columns without the hyperbole of its counterparts in Anaconda, Butte, and other cities, it did nothing to advance its readers' understanding of the points at issue in the conflict. Editorially, it was quiet on the subject when news of the president's involvement first broke, but after the "shock" had been somewhat absorbed, it printed an editorial titled "Sulphur Fumes." Stating that it had been "asked to explain" the smelter fumes problem, the paper merely repeated the Amalgamated version featured in the other papers. The whole matter was the product of an impetuous, anticorporation president, the *Tribune* implied. "According to the explanation of Amalgamated people," it said, the Tennessee example was not applicable to Montana because there was no "natural deposit suitable" for mixing with sulphur to make fertilizer and there was "no ready market" for the fertilizer. Consequently, its production was "altogether out of the question" at Anaconda. Anyway, the *Tribune* concluded, "The Amalgamated people had already studied the Tennessee plan for application to their own troubles, even before the president got busy."[8]

In concert with the campaign denouncing the administration's attitude on the smelter emissions, the papers engaged in a campaign against Roosevelt himself. The *Standard* set the style, characterizing the nature of Roosevelt's leadership as "the trouble in Washington" and posing the question, "Will He Be Good?" The president was admonished to settle down to "sane" conduct and stop "bickering with congress." His motives, the editors said, were based merely on an "impulse to bring Henry H. Rogers and others up with a round turn," an indefensible position, the paper explained, since it put thousands of "innocent men and women, citizens in peril."[9]

A not-so-subtle device employed by the papers was to reprint articles from eastern, anti-administration newspapers such as the *New York Sun* and the *Brooklyn Eagle*. The *Standard* reprinted an editorial from the *Eagle* titled "A Roosevelt Portrait" that pictured the president as an obstinate, self-righteous, narrow-minded tyrant: "The sum and substance of it is that no man, judge, politician, philosopher or plain citizen is safe, if his views are not identical with those of the president. To disagree with him is to be blind or cunning or law-defying or a wrongdoer or a swindler or a bribetaker or wrong-headed or traitor." The *Standard* supplemented the editorial with its own comment, assuring readers that "the *Eagle's* talk is typical of the run of current newspaper comment."[10]

The effect of the Montana protests on the Roosevelt men, for all the stridency, was almost imperceptible. The president's advisors were under no delusions about the reaction. It was what they had expected. The "Montana protests" were "altogether illusory," Attorney General Bonaparte told Roosevelt, summing up the administration's attitude. It was an old lament that had been encountered throughout the administration. To protest against the suit "was in itself a great outrage," Bonaparte said, but it was simply the familiar cry, "let us alone."[11]

The commotion was the predictable machinations of the trust when its purposes were questioned. "Merely to bring the suit and to proceed with it," the attorney general assured the president, "would not necessarily even disturb the business interests of the Amalgamated Company, unless, as has been suggested, *unless*, the company itself and those whom it controls purposely foment that impression and of their own notion choose to shut down." Bonaparte reaffirmed the prime objective of the suit: "What the Government is proposing to do on a question of general importance not restricted to this defendant and that locality, and on a good prima facie case, after careful examination, is to find out through the courts whether this defendant is doing what it ought not to do."[12]

William H. Taft acceded to the presidency in March 1909. He was attuned to Roosevelt's interest in conservation, and during the campaign of 1908 he had promised to convert his predecessor's executive policies into substantive legislation. Gifford Pinchot, chief forester for Roosevelt, had personally assured his supporters that the conservation policies of the Roosevelt men would be continued: "Taft is going to be with us in this work," Pinchot said. And, so far as the prosecution of the "Anaconda matter" and other smelters, Taft remained loyal to the promises of 1908.[13]

With Henry M. Hoyt, W. J. Hughes, and Ligon Johnson remaining in place in the Justice Department, a smooth transition in the case was accomplished. In order to facilitate the passage of needed legislation, Taft wanted a cabinet composed of the best legal minds in the country. For instance, to head the Justice Department he chose George W. Wickersham of New York, one of the most respected lawyers in the profession.[14]

Yet, even with the genuine commitment to codify the previous administration's policies, the Taft men had less confidence in the wisdom of federal jurisdiction and remedy at law than had the Roosevelt men. At bottom, the Taft men preferred to place more trust in the rectitude of honest private enterprise. Secretary of Interior Richard A. Ballinger perhaps represented this attitude to a greater degree than any other cabinet member. Wickersham's tendency along this line is reflected in an opinion he offered to Johnson regarding what he felt were the best means to bring about an enduring resolution to the fumes problem. Wickersham preferred, but not without some misgiving, private initiative to decrees of court. "I confess," he said in good earnest, "that I rely more upon the efforts being made at other places to eliminate the sulphur from the fumes by various processes, such as the one being practiced at Great Falls, than by a decree as the result of litigation."[15]

During the months of changeover in the government, constant pressure was applied to stop the inquiry. The pressure came from Sen. Thomas H. Carter and Sen. W. Murray Crane, an influential Massachusetts Republican who had easy access to Taft's counsel and whom Carter had enlisted in his attempts to stop the proposed suit.[16]

After a series of conferences with outgoing Attorney General Bonaparte and his aides about "the Anaconda smoke question," Wickersham came to the conclusion that the smoke case ranked as "one of the most important cases" under prosecution. But a number of complicated and hard-fought antitrust suits were also being carried forward at the same time. Most formidable among them was the effort to break up the Standard Oil trust, the financial interests of which—and many of its executives—were involved in the Amalgamated. On June 1, 1909, Wickersham, replying to a memo from President Taft, advised him that he had the Anaconda case "under careful advisement" and just as soon as he had received a report from an expert examining "the technical features" of the Anaconda plant he would "bring the matter before him."[17]

While the proceedings involved in the change of administrations were being completed, Special Assistant Ligon Johnson proceeded with his investigation. In his reports there was much his superiors could find encouraging, yet at the same time, problems began to surface that would become ever more difficult to contend with. In his investigation of the Washoe plant, Johnson had found particularly encouraging news. At the Washoe, he reported, "There were less obstacles to be found in converting the waste fumes of the smelters into sulphuric acid than were previously anticipated." Under the present means of operating the plant, the McDougal roasters, as previously understood by the Department of Justice, were responsible for as much as 60 percent of the fumes. And these fumes, Johnson said, were "practically pure so far as sulphur dioxide" was concerned. Therefore, there should be no problem in contending with impurities in the conversion of the sulphur fumes to sulphuric acid.[18]

But the principal reason for Johnson's optimism was that he thought the investment required to modify the McDougals would be slight; it was essentially "a mere matter of fume control presenting no difficulties to a competent chemical engineer." "No such change in method, or appliances, would be required to handle the Washoe fumes as was necessary in Tennessee," Johnson reported. "A slightly modified form of the McDougal plant as installed at the Washoe Smelter is erected in the various sulphuric acid plants of the country for the specific purpose of producing the very gas given off as a waste and destructive product at Anaconda."[19]

There was, however, one "chief question remaining" and one unknown variable. The unknown was the plans the company had for the internal operation of the plant. As previously established, next to the McDougals, the converter plant and then the reverberatories were the producers of the noxious fumes. But, Johnson claimed, these two departments "need not be considered in the effort to secure relief from present conditions . . . unless the company should see fit to transfer their operations from the McDougal's [sic] to the blast furnaces and reverberatories."[20]

The "chief remaining question" in regard to the McDougals, Johnson wrote, was "merely the cost of a plant to condense the fumes and the possible additional cost of installing additional McDougal roasters" to compensate for any loss in production. As a result of adding the modifications to produce sulphuric acid, the efficiency of the present furnaces might be reduced, Johnson explained. "To keep the plant at its present capacity it may be found necessary to erect one additional McDougal for every eight or ten now in operation, but beyond this no interference will be caused with the process or smelting results." To produce the desired results at Anaconda would require a plant of at least five times the capacity of the plant in Tennessee. To duplicate the single Tennessee plant would require about $600,000; thus Johnson concluded confidently, "From all the information at hand four million dollars appears to be a fair estimate of cost."[21]

This was a source of satisfaction to Justice Department officials. General counsel Kelley himself had established the capitalization of Amalgamated in replying to a formal interrogation by the department. The interrogation involved forty-one specific questions about the company, including the extent of its operations, the property it owned, and its financial condition. The "capital stock" of Amalgamated, Kelley wrote, was $155 million. The $4 million required to remedy the fume problem then was less than 3 percent of the capital available to the trust for operation of its business. This, department officials felt, was well within what was considered "fair and equitable to compel" the construction of fume-converting plants.[22]

Johnson's investigation into the commercial feasibility of the fertilizer market took him throughout the Midwest, Montana, and California. His findings on this question were no less encouraging. He found all the packinghouses and manufacturers of fertilizer materials, "other than those identified with so-called 'fertilizer trusts,' keenly alive to new possibilities" of acid production from both Tennessee and Anaconda. There was also lively interest in the prospect of opening the "recently discovered phosphate rock deposits." The manager in charge of the Cudahy Packing Company was "particularly courteous and willing to afford any information at his command," Johnson reported. However, the official did not want his comments made public. "Practically nothing," the official told him, was supplied the states between California and the Mississippi River, "there being no acid phosphate within this region available for mixing."[23]

Superphosphate would also find a market in "any locality wherein abattoirs" were located. It was mixed with by-products for fertilizer. Fertilizer among western farmers was virtually unknown, except in the form of manure or ground bone. They had endeavored to cope with soil exhaustion, Johnson said, by crop rotation. Several dealers in farm and orchard supplies were "of the opinion that an immediate market could be found if fertilizer could be sold in this section of the United States" for a "reasonable price." He could find no fertilizer dealers in Butte or Salt Lake, he pointedly reported.[24]

It was in Johnson's investigation into access to the phosphate beds in Idaho that discouraging signs began to appear. For every apparent solution, there developed a host of seeming problems. This was the beginning of an adversity that would dog government efforts for the next several years and would ultimately lead to accommodation with an uncomfortable truth. Writing from Los Angeles, Johnson told Wickersham, "The farther my examination extends, the more tracks of an Amalgamated cloven hoof I find."[25]

Since having left the East, Johnson had observed "a smooth shaven young fellow" who invariably turned up in all of his "round-about trips." The young fellow was spelled "occasionally by a thin, black-mustached chap. It looks as though someone is interested in my movements," Johnson observed.[26]

In California he conferred with officials of the Mountain Copper Company, which had proceeded under the threat of an injunction sought by the federal government a year earlier to convert sulphur fumes into sulphuric acid and then to merchandise the product through an auxiliary company by the name of the San Francisco Chemical Company. In order to secure suitable phosphate rock to mix with the acid, Mountain Copper had done extensive prospecting work in British Columbia, Mexico, and throughout the West. It was as a result of these investigations, the officials claimed, that the large phosphate beds in Idaho that extended into Wyoming were first discovered. The company filed claims and constructed a fertilizer plant at Martinez, California. The fumes from their copper were converted into acid in McDougal furnaces "practically identical with those at Anaconda," Johnson was informed. The phosphate rock shipped from Idaho was mixed with the acid and the product sold as fertilizer in California.[27]

Lately, however, there had been "agitation" over the rights of title to the phosphate beds: "Subsequent and antagonistic claims to the identical areas have been filed and considerable effort made to interfere with its manufacture of fertilizer," Johnson reported. He suspected agents of Amalgamated were behind the agitation: "From reliable information it appears that the Amalgamated Copper Company was well informed of this phosphate area for the past year or two. The experts employed by the Mountain Copper Company . . . proved to be thoroughly dishonest, and sold the information secured while in the employ of these interests to other parties, making trips to Anaconda among others in this connection." Other parties filing claims to the phosphate beds were the Stauffer Chemical Company and the Tennessee copper interests. Johnson requested authorization to "investigate in greater detail" the "schemes" that were apparently aimed at interfering with and invalidating the rights of the original discoverers. "The question will later be of considerable importance," he advised.[28]

Johnson's continuing investigation disclosed other complications with regard to utilization of the phosphate beds: "I find the Pacific coast smelters and also many owners of low grade copper ores deeply concerned in the disposition of these phosphate deposits," he reported. These interests were informed of the "proposed action against Amalgamated" and the plans to manufacture fertilizer. These interests, Johnson said, "would protest most bitterly any preference" that might be allowed the Amalgamated in gaining access to the phosphate lands.[29]

"Any preference of the Amalgamated will bring about complications and a most embarrassing situation," Johnson cautioned. It would probably compound the problems of resolving the fumes problem on a broad front: "Owners of low grade ores claim that only through the use of the by-products of the ore (the sulphuric acid), combined with the phosphate rock, to permit export and shipment throughout the United States, can their mines be operated profitably; and if the phosphate lands are open for lease or sale, a number of bidders may be expected."[30]

While in the West, Johnson, with the authorization of Solicitor General Hoyt, endeavored to reach an agreement with Amalgamated.

Fortified with his findings in the Midwest and California—in addition to the knowledge that the government had found a competent chemical engineer upon whom they thought they could rely—Johnson conferred with Kelley in Butte and Anaconda. Johnson wanted to test the "good faith" of Amalgamated, he said, in relation "particularly" to their agreement to cooperate with "any expert the Government might select." In addition, Johnson wanted to determine "whether or not any agreement which would in any manner bind the Amalgamated would be entered into by the Company." He thought it wiser to embody "any negotiations, promises or proposals in writing, and to request a written reply to anything submitted to the officers of the Amalgamated." He did this, he explained to Attorney General Wickersham, because "every verbal statement heretofore made by the Company's representatives invariably resulted in a misunderstanding when any performance was called for."[31]

In his report to Wickersham, Johnson forwarded a letter he had sent to Kelley earlier. Referring to any litigation that might be filed by the United States in seeking to secure the abatement of sulphur fume injuries, the letter had asked Kelley whether, "if any competent chemical engineer, after due investigation, should state it to be his professional opinion that it was practicable, not only scientifically but also commercially, to convert the sulphur fumes from the McDougals at the Anaconda smelters into sulphuric acid, you would abide the decision on that point." The government was willing to be as generous as it could possibly be and still protect its interest. Johnson had written Kelley: "The Government is perfectly willing that any litigation it may hereafter institute shall be suspended not only pending such an investigation but for a reasonable time thereafter in which to permit the installation of a sulphuric acid plant, should the report of the expert show that such a plant can be constructed."[32]

Before agreeing to make any reply, "Mr. Kelley insisted upon knowing just what expert would be designated." Johnson, sensing that he would get nothing without naming the expert, had given him the name. He had also agreed to an "arrangement whereby no public knowledge of any legal action would result, or any actual injury in any manner occur to his company." Johnson had then put the arrangement in writing, adding a concluding clause suggested by Kelley. It read, "This agreement is for the purpose of affording the parties opportunity of investigation in order that a settlement may be arrived at, if possible, without further litigation under the bill."[33]

Kelley took the letter and the "arrangement" to John Ryan. Several days later Johnson received a curt wire advising him that the company would not enter into the agreement. This, Johnson advised Wickersham, demonstrates "the uselessness of any further negotiations with the Company." The Amalgamated was, he said, determined to "fake things" and play for time until every advantage was theirs. It was evident "that the Amalgamated is prepared to fake things generally, under advice of an expert, before any government expert will be permitted to make an examination. Until every complication can be effected, no engineer representing the government will be welcomed at the plant."[34]

In view of the refusal by Amalgamated to enter into the agreement—after the government had conceded to all their conditions, and had even named their expert—Johnson next wrote W. J. Hughes, the Justice Department official in charge of the case, a letter stating, "I suppose Amalgamated are trying to find out whether or not it is possible to buy Mr. Nathaniel P. Pratt," the government's chemical engineer.[35]

Nathaniel Pratt was a chemical engineer, an admirer of Gifford Pinchot and of the administration's conservation program, and a successful sulphuric acid manufacturer from Atlanta, Georgia. He had worked with Ligon Johnson in the Tennessee cases. In late April 1909, Pratt and an associate, N. P. Heath, a chemist, conducted an examination of the Washoe plant on behalf of the federal government. The *Standard* reported their presence, referring to them ominously as "special agents" of the government. Their assignment, as written by Attorney General Wickersham, was to determine

> (1) Whether the gases escaping from the Roaster Plant [the McDougal furnaces] are sufficiently rich in sulphur dioxide to permit of their commercial conversion into sulphuric acid.
>
> (2) If . . . these gases should prove not sufficiently rich for that purpose, whether, by changes in the operation of the Roasters, the gases can be made sufficiently rich in sulphur dioxide to permit of their commercial conversion.
>
> (3) If it should be concluded that the gases are capable of being commercially made into sulphuric acid, what amount of sulphuric acid can be produced per day therefrom and, incidentally, what objectionable emanations from the roasters other than sulphur dioxide can be eliminated or converted into useful products.[36]

Pratt's report, comprising thirty-one typewritten pages plus graphs and photos, was received at the Justice Department in the middle of June. Its conclusions must have brought broad smiles to the faces of Wickersham and his aides, for the Pratt report vindicated arguments the government had been making for almost a year. The report concluded, succinctly, that

> The gases entering the final dust chamber from each of the eight individual sections comprising the Roaster Plant, are amply rich in sulphur dioxide to permit of their commercial conversion into sulphuric acid, and while they are not now to the ideal for this purpose, they are easily capable of enrichment to any degree desired for the manufacture of sulphuric acid without altering current methods of operation and without material decrease in the tonnage output required of each Roaster.

Although the entire roaster plant consisted of sixty-four McDougal furnaces, only thirty-nine were in operation during Pratt and Heath's investigation. Nevertheless, with only thirty-nine furnaces operating, 2,129.3 tons of chamber acid could be converted daily. And, the report added, "If sulphuric acid should be manufactured, it will dissolve and eliminate from the escaping gases a daily total of . . . 15.75 tons of white arsenic (AS_2O_3)."[37]

Officials felt they now had the information Roosevelt had earlier said had to be in place in order to determine whether a suit would be brought: "The entire question," he had said, was "whether the damage is being done, whether a fume converting plant is practicable even if costly, and whether it is fair and equitable to compel its construction."[38]

With the various investigations conducted by the Department of Agriculture in 1908 having established damage to the forest from sulphur dioxide; with Johnson's investigation having established the potential of the fertilizer market; with the figure of less than 3 percent investment required, an estimate based on Amalgamated's own figures; and with the Pratt report having established the scientific and commercial feasibility of converting fumes at the Washoe, Roosevelt's criteria had been met.

On March 16, 1910, James W. Freeman, U.S. attorney for the District of Montana, filed a suit in equity against the Anaconda Copper Mining Company and the Washoe Copper Company in the U.S. circuit court at Helena. The suit asked for a "writ of injunction properly restraining and enjoining" the Washoe smelter from liberating "any foul or dense or copper or sulphurous smoke, or any noxious, poisonous, unhealthy or disagreeable, or in any manner injurious vapor, gases, fume or odor" into the air or into "the waters therefrom." It also asked for further relief in whatever way the court might deem "meet and proper."[39]

The *Standard's* reaction was predictably outraged and immediate. The concern over the smoke had provided "abundant cause for intense anxiety once again." The federal government was making many "tens of thousands of good citizens . . . the victims of treatment that is getting to be nothing short of persecution." It assured the readers that as soon as Wickersham understood and appreciated "the spirit of the company in addressing itself to the possibilities" of remedy, the suit would be withdrawn. It was just another expression "of the universal agitation and an apparent inclination to hit a corporation head wherever it may appear."[40]

A month later, the Anaconda Copper Mining Company filed its demurrer, taking exception to all of the allegations of the government suit.

During the next several months the combatants girded themselves for the forthcoming court battle. "It looks to me," wrote R. L. Clinton to Johnson, "that they are preparing for an even more vigorous fight with the Government than they put up against the farmers." Profs. Theobald Smith, F. W. Traphagen, Joseph Blankenship, and Veranus Moore of Cornell, plus a battery of botanists and "Forest men," were "on the ground" for Amalgamated. A group of men were platting the entire area, and others were photographing it. After returning from a trip to the Anaconda region, Johnson remarked to Attorney General Wickersham that the company had so many experts in the field "they were tripping over each other."[41]

At the same time, the government was proceeding to strengthen its case. Profs. George F. Peirce, Robert. E. Swain, and J. Pearce Mitchell, chemists and botanists from Stanford University, and a plant pathologist were sent into the area to examine the condition of the country and secure evidence of the extent and cause of damage. Johnson secured photographs of the smoke stream going over the national forests from an "A-1 photographer" in Butte who could be "absolutely relied upon to keep his mouth shut."[42]

Johnson also began tracing prior testimony given by Amalgamated experts in other suits involving similar injuries, especially the testimony of Profs. F. W. Traphagen and Frederick J. Falding. Falding, Johnson said, had given some "very far fetched testimony" in the Bliss case with respect to his "reputation as an expert" and as to the "impracticability of changing the method of operation" of the Washoe plant. "The Anaconda people," he surmised, "will probably rely heavily upon Falding's testimony as to cost of the sulphuric acid plant, his plan of the plant and otherwise." The New York papers had also run a story on Falding's divorce proceedings, which, Johnson claimed, placed his "standing, earning capacity and the like" in question. When the case was tried, Johnson told W. J. Hughes, "I do not want to leave any possible loopholes." He expected the testimony the company experts had given in previous cases to "discredit to a large degree the evidence they will give in our suit."[43]

As the contestants were busy fortifying their positions legally, scientifically, and technologically, action was proceeding on the political front. Joseph M. Dixon's initial response in 1908 to the threat of the injunction was that "one week's wages at the Anaconda smelter and in the Butte mines is of more value than all the lodgepole pine affected by the smelter smoke." After conferring with Roosevelt, Dixon had released a statement assuring Montana voters that the government did not intend to act capriciously: "It was not the purpose of the government to take action that would close the smelters in case no reasonable or practical method could be discovered for taking sulphur and arsenic out of the smoke," he said. "The government does not desire to inflict hardship by imposing impossible conditions, but will insist that every reasonable effort will be made to devise a practical method of desulphurization." Thereafter he stood aloof from the controversy and allowed it to take its course.[44]

Senators Carter and Crane, however, endeavored to use their influence to persuade the government to terminate the proceedings. Despite Senator Carter's efforts, Attorney General Wickersham told the senator that he intended to pursue the suit "with vigor." Carter's colleague, Senator Crane, still thought his influence would carry the day. Answering one of Senator Carter's letters, he confidently wrote, "Your letter of the 8th is received. I will write to the Pres. or to Mr. [Charles D.] Norton [Taft's secretary], who will call the matter to the president's attention, today urging him to do as you wish regarding the smelter smoke situation, and I have no doubt that he will take action."[45]

Throughout the controversy, Carter tapped popular anxieties while serving the trust. While he was senator, he assured voters, the government's suit would not be pressed. His statement mollified the people of the mining districts, but it enraged the Deer Lodge Valley Farmers' Association.[46]

Late in 1910 the Justice Department received the two reports that, coupled with the Pratt report, would furnish the basis for their arguments against Amalgamated's experts. George Grant Hedgcock, forest pathologist for the Department of Agriculture, completed his study, "Report on Forest Conditions in the Vicinity of Anaconda, Montana, as Compared with Those in Adjacent Forests," and Professors Peirce, Swain, and Mitchell of Stanford University completed their investigation of the area surrounding the Washoe smelter.

The Hedgcock report, as its title indicates, was a comparative study of the conditions of the national forests in Montana and Idaho. In Montana, the Absaroka, Gallatin, Madison, Beaverhead, Deerlodge, Missoula, Bitterroot, Flathead, and Blackfoot (Glacier National Park) National Forests were studied; in Idaho, the Targhee, Wiesel, and Salmon were studied. In addition, the condition of adjoining private forests was analyzed. Special attention was paid to the general health of the trees, the prevalence or absence of fungous diseases and injurious insects, and the extent of injuries by fire and winterkill. In the Deerlodge National Forest and in private forests within a twenty-five mile radius of Anaconda, particular attention was devoted to the growth and general condition of the trees as compared with the growth and condition of trees in adjacent and more remote forests in Montana.[47]

Hedgcock came to the definite conclusion that a very large percentage of the timber within a radius of twenty-two miles of Anaconda—with the exception of a few species of no commercial value that were peculiarly invulnerable to fume injury—was being killed by smelter fumes and that the nature of the damage was not similar to dying timber in other localities adjacent to and distant from the Anaconda region. He also showed that the timber that did survive within the twenty-two-mile radius was being retarded in growth, even though not dying. In general, as the distance from Anaconda increased, the amount of living foliage on the affected trees increased. As a result of the irregular pattern of the smoke column, the timber was not uniformly injured or retarded in growth. Trees outside the Anaconda region "differed radically" from those inside the area possessing, in general, "twice the amount of foliage."[48]

The study of Peirce, Swain, and Mitchell had a twofold purpose: (1) to establish "scientifically the scope of actual fume destruction upon Government lands," and (2) to establish the "scope of fume influence preventing maturity of plants, particularly forest trees." Joining them in conducting the investigation and gathering data were J. K. Haywood, who had made the earlier studies of fume damage in the Deer Lodge area and in California; Professor Harkins, who had made studies on behalf of the farmers in the Bliss case; and three other scientists from the Department of Agriculture.[49]

Photographs of the hills south of the smelter compared with photos taken three years earlier showed that the "bareness" of the hills had increased and the few trees that had been alive were now dead. The turf on the sides of the hill had increased and erosion was more noticeable. The "heavy forest, lumbered off for the most part," they observed, was not reproducing itself. "There are neither saplings nor seedlings, and now that the seed trees are also gone, natural reforesting would be very slow, even if smelting ceased or were so carried on that no poisonous wastes escape. Injury has been done which will endure long after its cause ceases."[50]

To the west, in sheltered areas and along Warm Springs Creek, "the beneficial effects of sufficient water on plants struggling against any unfavorable influence is plainly shown," the investigators reported. In such areas, "even within five or six miles of the smelter," trees appeared to be thriving. Writing as if they were lecturing an Introduction to Forestry class, they explained, "But cutting these trees, which look young because they are short, reveals the fact that they are really old, their annual rings being very narrow. These rings, being the annual addition to the thickness of wood in stem and branches, plainly indicate not only the age but the favorableness or unfavorableness of conditions during the growing seasons of a succession of years." One fir tree, only four feet high, was cut at ground level. Its age was found to be at least twenty-five years.[51]

Numerous observations and samples of this sort were made and catalogued—as were measurements of the annual increase in length of needles, counts of the needles retained year to year, specimens for laboratory examination, and photographs taken of the area in a radius of up to ten miles from the smelter. Wherever the readings were taken, the results were generally the same—damage had intensified. "Even when smoke may not hurt vegetation sufficiently to show as spots and burns on the leaves or as death among the plants most susceptible," the scientists stressed, "it may nevertheless be recognized as injurious by the thinness of the foliage and the location of the leaves."[52]

They charted and chased the smoke stream around the region, obtaining "air-analyses" at various points and distances from the smelter. "Only in this way," they said,

> could we connect the damage everywhere evident with the smoke as a cause, and only by ascertaining that the smoke is present in the

> field in concentrations which our experiments at Stanford University have shown to be injurious to forest trees and other plants.
>
> At a distance of five miles southwest from the stack we were in smoke so dense as to make the sunshine seem pale despite the fact that the day was remarkably clear outside of this valley. . . . Air-analyses showed that this vicinity had been subjected for hours on this day to a bath of at least double the concentration of SO_2 which we had found by experiments to be injurious.[53]

At some locations they saw evidence that other studies had been conducted on the timber. At one such place, they conducted their study with special care, because they had been informed that Professor Traphagen had been working in the area. They noted that "the tips of branches, especially lodge pole pines, had been cut off clean with a knife." It was important to note, they told Johnson, "that in every case we observed the tips of the branches only had been taken from the field; they are the least affected part of the tree and have the lowest sulphur content." Their overall conclusion was that in spite of the difference in altitude, "and the other differences attendant on this," the cause of injury and death in the trees was the relative impurity of the air.[54]

Throughout the latter months of 1910, talks went on between Johnson and Kelley in New York. The government was still hoping to resolve the problem in the manner initially envisioned by Theodore Roosevelt. The Taft men wanted a workable settlement that would obviate a grueling court battle and a hostile relationship with an industry they knew could devise the means to correct a problem for which it was responsible. In October, when the government's investigation was at its most active, Cornelius F. Kelley began talking in terms that promised realization of this goal. Writing to Johnson about control of stack emissions, he espoused what appeared to be everything the Roosevelt and Taft administration had hoped for: "I assure you that it is the intention of this company," Kelley said, "to keep in the foreground in this matter, and that it will not be slow to adopt any measure which is practical and which promises satisfactory results."[55]

Johnson and Wickersham were guardedly optimistic, and they set about preparing a "general proposition of settlement" to submit to the company's board of directors. An effective settlement appeared imminent.[56]

– Chapter Eleven –

"We Will Have Secured All That We Have a Right to Ask"

Early in December 1910, after a number of talks with Ligon Johnson, special assistant to the attorney general, Cornelius F. Kelley, general counsel for Amalgamated Copper, submitted a "draft of proposition" for use as a basis for negotiations. Kelley argued that, should an attempt be made to arbitrate the damage, the company's experts and the government experts would be "hopelessly at odds." He did not consider the effects of erosion as a part of damages, which Johnson contended "quickly renders mountain soil so light that vegetable life is practically prohibited." Kelley thought "the Government should be content to accept a minimum sum under stipulations" that the Anaconda Company "use all possible efforts to abate the cause of injury." Compensation should be fixed upon the value of standing timber and only upon timber killed within the last seven years. The stumpage price should be figured, Kelley claimed, "at a price less than one half that the Government now gets." He offered to pay a twenty-five thousand dollar lump-sum settlement for past injuries and five thousand dollars yearly "for continuing damage."[1] Amalgamated's president, John D. Ryan, pledged that in exchange for this agreement the company would "agree to diligently carry on investigations with the object in view of preventing sulphurous fumes from escaping from our stack." In turn, Johnson recommended to George W. Wickersham, the attorney general, that the government ask for "annual payments so large that they would guarantee active investigations and efforts by the Company to remedy the smoke situation."[2]

Replying to Johnson about Kelley's proposal for settlement, Wickersham considered the attorney's figures "absurd" and the stipulations meaningless. His terse reply indicated he was not intimidated by the thought of confronting the Amalgamated's experts in arbitration proceedings or in the courtroom: "If we do not get any nearer to a real settlement than indicated here," Wickersham wrote, "there will be nothing for the Government to do but to go on with the law suit." Johnson agreed but thought they should wait the Anaconda officials out, for "If any reasonable proposition is made, it will not be submitted until the last minute, and then only when the officers feel quite sure that every other scheme has been exhausted." They would also, Johnson thought, want to include the "McCune cutting" case in any proposition for settlement.[3]

The McCune cutting was a timber depredation case that had occurred several years before. It was a relatively small area within the Deerlodge National Forest that the Anaconda Copper Mining Company had logged without reimbursing or seeking approval from the national government. Furthermore, because of the fumes, the area had failed to reforest. But Wickersham refused to consider the McCune cutting within the settlement of the fumes problem.[4]

Within weeks of having made his first offer late in 1910, Kelley submitted another draft of terms the Amalgamated considered appropriate for negotiating settlement. Wickersham saw little improvement and told Johnson, "I do not know that I have anything to say to Mr. Kelley unless he shall submit some feasible proposition. I do not propose to get into a general discussion with him with no basis on which to work. I do not feel that his last proposition is sufficiently tangible to make that the basis of any discussion."[5]

Talks continued through January and February 1911. By late March the parties had narrowed the points of discussion to the ambiguities of language in the proposed settlement. General counsel Kelley wanted to be certain, said Wickersham, that "no possible injustice" could be done his company.[6] During the course of the talks, no action was taken on the demurrer filed by the mining company. Then on April 13, 1911, in Washington, B. B. Thayer, president of the Anaconda Copper Mining Company, and George W. Wickersham on behalf of the United States of America solemnly agreed to a stipulation. On the same date the trustees of the Anaconda Copper Mining Company agreed to a resolution affirming the terms of the stipulation.[7]

The heart of the contract read as follows:

> The defendant Anaconda Copper Mining Company agrees that it will at all times use its best efforts to prevent, minimize and ultimately to completely eliminate the emission and distribution from its smelting works at Anaconda, Montana, of all deleterious fumes, particularly those containing sulphur dioxide. But the said defendant by entering into this stipulation does not concede that it has caused to be emitted from such works any fumes which are injurious to any of the interests of the complainant.

To assure compliance with the stipulation, a board of experts was established:

> For the purpose of determining whether or not the said defendant is at all times complying with the foregoing stipulation, it is agreed that Messrs. John Hays Hammond, as chairman, and J. A. Holmes and Louis D. Ricketts are hereby appointed a Board of Experts, which Board shall from time to time investigate and report to the parties hereto concerning the operations of the said defendant's plant with respect to the emission of fumes containing deleterious

> matter injurious to the interests of the complainant, and with respect to the methods, devices or processes adopted by the said defendant to comply with this stipulation.

The mining company agreed to cooperate with the board and the attorney general in carrying out the terms of the agreement:

> The said defendant agrees from time to time to conduct its operations and introduce such methods, improvements, devices or processes as such Board, or a majority of its members, shall from time to time in writing certify (1) to be best calculated to carry out the provisions of this stipulation, and (2) to conform to the most scientific processes then practically applicable to works of the character of those of the said defendant. The said defendant agrees promptly to comply with all requirements pursuant to such certificate, as and when so certified, and if and when requested by the Attorney General of the United States.

Provision was made for continuously filling vacancies on the board created by the death or resignation of any member.[8] Other provisions provided for compensation for the board and the employment of experts to assist the board.[9] The stipulation concluded with terms providing for remedy in the event of a breach of the contract: "So long as the said defendant shall well and truly comply with the terms of this stipulation, no further proceedings shall be taken in this suit. In case of a breach of the terms of this agreement then the Government may apply to the court for such relief as it would be entitled to if this agreement had not been made."[10]

Attorney General Wickersham accepted the stipulation in lieu of a lump-sum settlement and annual payment for damages because he reasoned it would be better to wait until "all harmful agents were removed" and a "complete cessation of injury" was reported or established before taking up the question of damage. Writing to the solicitor general in 1914, Ligon Johnson explained Wickersham's position regarding final settlement and the question of compensatory damages. Referring to the lump sum offered by Anaconda Copper, Johnson observed, "This the Attorney General refused, inasmuch as it was not only considered inadequate but also was thought to jeopardize the Government's rights in the event it became necessary to proceed further under the bill. The question of damages was left to such time as the Commission should report the damage practically abated."[11]

In view of the terms of the stipulation and the caliber of the men making up the board of experts, which, throughout the industry, soon came to be called the "Anaconda smoke commission," both Wickersham and Johnson thought the agreement was the best they could expect to achieve and considered the smelter fumes problem to be on a course to resolution. Writing to Johnson after all the parties had executed the stipulation, Wickersham commented, "I have communicated with all the members of the Commission: they are all interested in the problem, and I think this will work out to the great advantage of the Government. At all events, we will have secured all that we have a right to ask."

Johnson expressed full agreement: "I think the Government is to be congratulated upon securing such an able commission. I have no suggestions to make as to any change in the stipulation."[12]

The two men had presupposed that the interests of the trust and its politics would stop at the commission's doorway. Viewing their accomplishment from the perspective of 1911, devotees of good government could conclude the war had been won. All that remained was the enforcement of the provisions. Waging the peace, however, would prove far more difficult than waging the war.

The men appointed to the commission were, indeed, men of a high caliber. As mining engineers and entrepreneurs, John Hays Hammond and Louis D. Ricketts were the best the mining industry could produce. Chairman Hammond was a towering figure in the industry as well as in American society. A graduate of Yale and a close personal friend of President Taft's, he played a prominent role in Republican Party affairs, once being mentioned as a vice-presidential candidate. Taft appointed him special ambassador to represent him at the coronation of King George V in 1911.

Hammond was also something of a soldier of fortune, having been associated with Cecil Rhodes in the development of gold mines in South Africa in 1895–96 and one of four leaders in the so-called Transvaal reform movement. Although out of sympathy with the abortive Jameson raid, a cause of the 1899–1902 Boer War, he had participated in it, been arrested, tried, and sentenced to death. The sentence was commuted to fifteen years' imprisonment, but later, upon payment of $125,000, he was released. He furthered his mining engineering training at the Royal School of Mines in Saxony and took a doctor of laws degree at St. John's College in 1907. He became associated with some of the most important financial groups in the United States, purchasing and promoting several of the largest and most valuable mining properties in this country and in Mexico. At the time of his appointment to the commission Hammond was reportedly working "independently," but prior to this he had served as a consulting engineer to the Guggenheim mining interest at a salary reported to be $250,000 a year. He was, said the *Standard* proudly, "acknowledged as the foremost mining engineer in the world."[13]

Louis D. Ricketts's public appearance was the antithesis of the illustrious Hammond's. Rarely was he seen in a business suit, preferring instead the traditional garb of the mining engineer, the soiled slouch hat and rumpled khaki shirt and trousers. At the Greene-Cananea mining camp in Mexico he was called *mal cintado*—"the badly belted one"—by the workers, since his pants always appeared on the verge of falling down. But he was perhaps the foremost expert on the construction of copper reduction plants. One writer claimed that "no one ever rivaled him in the number of large and successful plants for the reduction of copper ore he built." He had been employed as a consulting engineer for a number of copper mining companies, most notably Phelps Dodge and Company. At the time of his appointment, he was a director of that company and also general manager of the Greene-Cananea Mining and Smelting Company, a part of the Anaconda mining empire since 1906. In addition, he served as trustee of the California Institute of Technology.[14]

J. A. Holmes was director of the U.S. Bureau of Mines. The bureau, a unit of the Department of Interior, had been established to provide means for eliminating health hazards, promoting safety in mines, and to provide statistical information on domestic and foreign mineral production and consumption. Through the years, however, its concerns dovetailed with the interests of the mining industry, and for the most part, it served the industry through its production-profit-protection orientation. This orientation, unfortunate but inevitable as it was, had evolved from the fact that the careers of the professionals in the Bureau of Mines were inextricably tied to the various mining companies. Many in the bureau in the early decades of the twentieth century moved back and forth between industry and government service, becoming in the process disposed toward the mining and smelting industry.[15]

This is not to imply, however, that J. A. Holmes himself did not have a genuine interest in and commitment to remedying the emissions of smelters in the fashion the government pictured. Most of Holmes's professional career had been spent in government service, first with the U.S. Geological Survey and then with the Bureau of Mines. Wickersham and Holmes had a number of talks regarding the emissions problem. Following one talk after the commission began its work late in December 1911, Wickersham observed to Johnson, "I think that [Holmes] has the matter well in hand and I am well satisfied that the Government's interests are being satisfactorily looked after."[16]

Under Holmes, an office of air pollution was established in the bureau and research begun into means of controlling excessive smoke emissions. The efforts of the new office culminated in 1915 in a comprehensive article on smelter emissions, "Metallurgical Smoke," written by Charles H. Fulton. Holmes died in 1916 and the office of air pollution was abolished shortly afterwards. It may have been abolished for the reason posited in Fulton's article.[17]

Fulton contended that the problem of smelter emissions would only increase, and the industry should therefore develop "smoke engineers" to deal with the problem. The emphasis in his article was on means of recovering wastes in the smelting process. He maintained that smelting plants were "making every effort to devise ways and means to do away with possible damage and annoyance from smoke and meeting with success," but added that "the solution of the problem is not yet at hand and much work still remains to be done." Then, bowing to the imperative of "the magnitude of interest," he concluded: "As the mineral industry is one of the great basic industries of the country and of necessity is entitled to full consideration, it should be accorded freedom to work out the smoke problem to the benefit of all concerned." The agency by which this freedom should be accorded, he stated, was through an "independent commission like that of the Anaconda Smoke Commission."[18]

During the proceedings and negotiations with Anaconda Copper, the Department of Justice had been reluctant to initiate legal action against the offending smelters in California. During negotiations, Wickersham directed "that actual legal proceedings be deferred [in California] until the autumn, and that, during the summer, a further survey be made by the Interior Department of the situation and report on the condition then found to exist."[19]

Investigations in California had been initiated in 1908 by Attorney General Charles J. Bonaparte, and by early 1911, bills of injunction had been prepared and evidence worked up for trial. At that time, Secretary of the Interior Walter L. Fisher arranged for J. B. Chatterton of the General Land Office to conduct a confidential inspection of the area around the smelters situated in a deep narrow canyon about thirteen miles north of Redding, California, on the Sacramento River and its tributary, the Pitt. Investigations made earlier by Johnson and J. K. Haywood had revealed that, although the area around the smelters was once heavily wooded, by 1910 it was "swept bare of vegetation for miles." "As far as the canyon itself is concerned probably all the damage possible has already been done unless reforestation were undertaken," that report read. "The latter event would probably be slow and difficult work since the loss of vegetation [has caused] the steep hill sides . . . [to be] washed bare of soil for miles around."[20]

And, the investigators reported, the smelters were "operating near enough to the forest reserves and public domain as to threaten widespread damage." Farmers situated south of Redding, where the canyon widened out into the fertile Sacramento valley, formed the Shasta County Farmers' Protective Association and sought to enjoin the smelters from operating because of increasing damage to crops.[21]

The conflict in Shasta County developed along the same lines as the conflict in Montana, except for the decision rendered in court. In the beginning, a series of meetings were held between the farmers' association and representatives of the smelters, the farmers demanding the emissions be controlled and the smelters' counsels demanding more time to "consider" the matter. Complaining to a representative in Washington, the association's chairman, Chub Gibson, wrote, "The last meeting was a scene of wild confusion. The lawyer representing the Mammoth Copper Company told the members that evening that they were a 'flock of cackling geese' and didn't know enough to realize whether the smelter fumes were doing any damage or not." Describing their region as "literally scorched,"

Gibson pleaded for assistance: "Our only salvation lies in Federal aid, and I humbly ask you for the same." Writing to the Department of Agriculture, he pointed to a paradox: "The Government of the United States is spending millions of dollars annually establishing irrigation works in order to furnish water for present and future farms. The people of the fruit-growing and farming districts of Shasta County 'humbly ask' that the Government make all smelter companies in this section of the country control their smoke, in order to save the fruit and farming industries already established."[22]

Meanwhile, at Stanford University, as negotiations with Anaconda representatives proceeded and the farmers fought with the smelters, Profs. George F. Peirce, R. E. Swain, and J. Pearce Mitchell continued their studies on the effects of smelter fumes on plant life and the extent of damage. Their laboratory work confirmed their initial field observations: "We have collated recently the results of the chemical analysis of most of the samples collected by us last summer in the vicinity of Anaconda," they reported to Attorney General Wickersham in 1911. "These results confirm our field observations in a most decisive way, both in regard to the character and the extent of the injury to the forests of that region." Furthermore, they continued, "When the results are compared with the data secured by Dr. J. K. Haywood of the Department of Agriculture in his investigation of that district in 1908, the fact that the damage has progressed greatly in the interim is very clearly shown in the new limits of injury which they help establish."[23]

On the "botanical side" of the study they had established the nature of the action of sulphur dioxide (SO_2) upon the cells and tissues of plants:

> A study of these sections proves that the appearances in the cells and tissues are not produced by fungi, bacteria, or insects, but are due to a poison absorbed in gaseous condition. This microscopic study accounts for the decreased growth in smoked plants as

> compared with unsmoked ones. Our experiments with pure SO_2 show what this poison is, and that it interferes with the normal activity of the food-manufacturing apparatus of the plant.

Their experiments had not yet determined the lethal dose of sulphur dioxide for any plant but with continued experiments they expected to establish it. Johnson, looking to the approaching litigation in Montana and California, urged the Justice Department to fund the remaining part of the study because he wanted to use their studies as a "scientific basis for all future action." The professors volunteered their services and asked for $250 to acquire "the necessary laboratory assistance."[24]

After achieving the stipulation with Anaconda, the Justice Department moved against the California smelters that were damaging forest reserves: the Balaklala Consolidated Copper Company (First National Copper Company); the Mammoth Copper Company, a subsidiary of the United States Smelting and Refining Company; and the Bully Hill Smelter owned by the General Electric Company. Suits in equity and bills of injunction were prepared, but before they were filed, conferences led to agreements between the United States and each of the smelters.[25]

Earlier, the Shasta County Farmers' Protective Association had instituted suit against the Mammoth and the Balaklala smelters and in 1910 won a decision against them in the U.S. Circuit Court of Appeals. The decision in the case followed closely the decision granted in Utah in 1906 in the case of James Godfrey et al., representing about four hundred farmers, versus the American Smelting and Refining Company, the United States Smelting Company, the Bingham Consolidated Company, and the Utah Consolidated Company. Utah Consolidated was part of the Amalgamated's industrial family. The plants operated around Salt Lake City. The Bureau of Mines reported in 1915 that air containing 0.007 percent by volume of sulphur dioxide was "unbreathable" and caused "a spasmodic contraction of the air cells of the lungs." Choosing a number of industrial cities throughout the world suffering from air pollution but neglecting to identify the period, the Bureau of Mines reported that Salt Lake City's air contained up to 0.001 percent by volume of sulphur dioxide, a far larger amount than any of the other cities listed.[26]

The Utah decision was designed to prevent any of the companies from treating ore containing more than 10 percent sulphur in the natural state and from allowing the escape of arsenic from the stacks. By permission of the court, the American Smelting and Refining Company and the United States Smelting Company entered into agreements with the farmers whereby their flue and dust chambers were revised and "baghouses" constructed to filter the smoke stream through woolen or cotton bags.

The Utah agreements were comprehensive, calling for the fulfillment of four conditions: (1) all free sulphuric acid had to be removed from smelter smoke; (2) all solids had to be removed from the smoke by means of a baghouse; (3) after the removal of solids and sulphuric acid, the gases discharged into the atmosphere had to contain not more than 0.75 percent of sulphur dioxide; and (4) the sulphur dioxide that escaped must not damage crops and livestock or cause discomfort or inconvenience to people. The Bingham Consolidated Company and the Utah Consolidated Company closed their smelters, moving their operations to Toole, Utah, about twenty miles southwest of Salt Lake City. It was a region in which "smoke can seemingly do less damage than in the immediate vicinity of Salt Lake City," a Bureau of Mines spokesman commented, yet, he added, "it is doubtful whether the matter is permanently settled."[27]

The agreements reached with the Justice Department by the Balaklala and Mammoth in California called for the installation of devices to eliminate poisonous fumes. The Mammoth installed a baghouse and instituted a number of supplementary processes. Ligon Johnson wrote the attorney general that the method "appears to neutralize the killing effects of sulphur fumes from ores."[28]

Professor F. G. Cottrell, renowned chemist at the University of California wrote in the *Journal of Industrial and Engineering Chemistry* that the Mammoth's smelting process represented a "notable achievement, being the first time that the bag house, so efficient in lead smelters, has been successfully applied to copper blast furnace gases on the larger scale. It is made possible in this instance through neutralization of the sulphuric acid in the gases by the zinc oxide carried over in the fume from the very heavy zinc content of the ore smelted."[29]

The Balaklala stipulated to install the Cottrell electric precipitation method at its plant in California. Developed by Professor Cottrell, the method used an electric current at high potential to precipitate suspended particles from the smoke stream. The air surrounding the electrodes was charged with electricity and the solids in the smoke were attracted to the electrode plates of opposite charge. The process "worked splendidly when the suspended particles were wholly solids or wholly liquids," Johnson reported to Wickersham, "but when both liquids and solids were to be precipitated the plates would become clogged." "Considerable difficulty was found in cleaning the plates," Johnson reported. "The liquids and flue dust would in effect form a paste which would harden and put the plates out of commission. At best, when shaking of the electrodes did not clean the plates it was necessary to temporarily divert the gas in an untreated state."[30]

When the Balaklala was unable to comply with the agreement, the farmers sought an injunction. At the hearing it was shown the company was not removing all solids and that sulphur dioxide emissions ran about 0.075 percent. The court granted an injunction on July 25, 1911, and the plant closed.[31]

In its agreement with the Justice Department, the Bully Hill took a different approach: It agreed to close down their Mammoth plant, among others, during the growing months of June, July, and August "until some other company had developed a process which they felt willing to adopt at their plant." They were "perfectly willing," Johnson told the attorney general, "to close if any excuse could be offered and they preferred to do this, in as much as they were making nothing under their then operations."[32]

After obtaining the agreements with the California smelters, the Department of Justice conducted an aggressive campaign within the smelting industry to develop metallurgical methods that would not only extract the metals from the ore but would also prevent the production of injurious emissions. Experiments were conducted and studied in all three classifications for chemically processing ores: pyrometallurgy, hydrometallurgy, and electrometallurgy. An argument made by some experts of the industry, particularly the copper industry, was that the sole source of injury to vegetation was from suspended solids and sulphur trioxide. However, Dr. Peirce's extensive studies at Stanford "exclusively demonstrated," so far as the Department of Justice was concerned, that sulphur dioxide alone destroyed vegetation. Therefore, their chief objective was to discover or identify a metallurgical process that would abate this gas. Their investigations of the baghouse at the Mammoth smelter and the Cottrell electrical precipitation method at Balaklala convinced them that solids could be removed from the smoke stream successfully without undue commercial hardship.[33]

In addition to the baghouse and Cottrell methods, numerous other processes and theories were investigated. Among them were methods described as concentration and leaching, straining through water or spray, the Westby-Sorenson smoke process, fume treatment with milk of lime, superheated steam with furnace smelting, and the Reed combined smelting and leaching process. In Salt Lake City the investigators even encountered a disciple, presumably, of "Old Hutch," the "smoke messiah" of Butte. Thomas W. Lambert, described by Johnson testily as the "most persistent advocate of a remedy," wrote "volumes to the department in advocacy of his plan" to "destroy" the sulphur via his incinerating furnaces. After several months of hearing technically bewildering chemical and metallurgical testimony by a variety of experts on the pros and cons, virtues and evils, potential and nonsense of the various methods and devices, the department settled on the one process they felt warranted further examination.[34]

The Thiogen process, originated by Dr. Stewart W. Young, had been "fully developed by demonstrations on a small scale," Johnson told Wickersham optimistically: "The commercial feasibility of the process in conjunction with smelting on a large scale has yet to be proven. Dr. Young has associated with him a prominent chemical engineer who assures me that all engineering difficulties have either been overcome or are easy to meet. The process has been duly financed and preparations are under way for the erection of sample furnace and apparatus to demonstrate the process."[35]

Arrangements were made for a test of the Thiogen process as well as a number of others at Los Angeles and at Palo Alto, California. The representatives of "interested smelters," the Anaconda smoke commission, and the Bureau of Mines were urged to attend by Attorney General Wickersham. At the demonstration were members of various technical journals and chemists from the Amalgamated plants at Great Falls and Anaconda and from Utah and California smelters. Professor Cottrell, now with the Bureau of Mines, represented the Anaconda smoke commission.[36]

Dr. Cottrell prepared a comprehensive report of the Thiogen process for the Anaconda commission. The process provided for the direct conversion of fumes into commercial sulphur. Reporting to the attorney general on the results of the test, Johnson wrote that it was carried out under laboratory conditions but on an "extensive scale." The test "proved conclusively that the Thiogen process was chemically feasible." One smelter, that of the Penn Smelter Company, near Valley Springs, California, began installing the process in their operation even though all of the mechanical and engineering questions reportedly had not been resolved. Johnson believed the process would obviate the economic arguments of the smelters if the Penn's venture succeeded. He believed that

> If the mechanical and engineering difficulties are successfully met in this installation, and if the cost of operation can be kept anywhere near the estimate of the inventor, this process will afford a commercially feasible plan for smelter operation under which sulphur dioxide injuries may be eliminated.
>
> It would appear that there is certainly good grounds to hope that the installation of the process at the Penn smelter will afford ample precedent for the Government in the Anaconda Case.[37]

By December of 1912 the managing director of the Penn smelter informed Johnson that the Thiogen process had been installed and the "operation of the smelter reducing plant [was] successful." He reported further that "All the engineering difficulties have been overcome and the engineers are now busy overcoming minor defects and making such changes in the apparatus as actual operation demonstrates to be necessary." The best news of all, however, was that the sulphur could be marketed to Standard Oil: "I have learned confidentially that the owners of the Thiogen process," Johnson wrote Wickersham, "have practically completed an arrangement with the Penn Salt Company which is, in effect, Standard Oil, under which the gross sulphur output will be taken over by this company at a price sufficient to cover the cost of operation of the Thiogen plant with an element of profit in addition."[38]

Another process that the Justice Department felt showed promise of "curing the sulphur dioxide problem" was the Hall process. Invented by William A. Hall, it was designed, as was the Thiogen process, to remove sulphur directly from sulphide ores in the form of elemental sulphur, hence preventing the formation of sulphur dioxide. It seemed to be ideally suited to the Washoe smelter operation. Writing in 1915, Charles Fulton of the Bureau of Mines explained that the process was being applied to the roasting of sulphide ores in furnaces like the McDougal—the chief source of sulphur dioxide fumes at the Washoe smelter. The Balaklala was using the process "on a large scale" in their plant, Fulton reported, but did not assess its performance.[39]

The Bully Hill had remained closed, sending its ores to other smelters located more favorably in relation to the agricultural interests. It was, nevertheless, conducting hydrometallurgical experiments, looking toward the extraction of zinc and copper by leaching and thereby avoiding the fumes problem.[40]

In the meantime, the Department of Justice had received the confidential report on forest conditions around the three California smelters. The report concluded that conditions on the public lands in Shasta County in 1912 were no worse than when they had been investigated in 1909. The reason was simple enough: "It appears," deduced the investigator, "that no further damage to timber in the vicinity of these smelters is possible as all available to the fumes are destroyed."[41]

In late 1912 Ligon Johnson concluded that unless the smelters changed their methods of operation, nothing remained "in these matters but the question of compensation for damage done before any bills for injunction were prepared." Within a year, settlement was reached by calculating damages on a stumpage value—the value of the timber as it stands uncut. With practically no argument, the Mammoth paid the United States $2,520.75 and the Balaklala paid $1,263.00. No payment was asked of the Bully Hill because it was agreed it had not operated long enough for its fumes to have damaged timber beyond recovery.[42]

Meanwhile, remedy of the Anaconda problem appeared close. "There is little to be done in the Anaconda Case beyond thoroughly testing each new fume device offered," concluded the Justice Department. And the most promising process, under the pen of Dr. Cottrell, was "rapidly getting into form for final consideration by the Committee."[43]

– Chapter Twelve –

The Struggle Abandoned

"Whatever improvements at the Washoe Smelter may be developed as a result of these inquiries and investigations will be highly beneficial to the metal mining and metallurgical industries of the entire county," wrote Chairman John Hays Hammond of the Anaconda smoke commission, revealing the narrow corporate interests of the commission members in his first report to the U.S. attorney general in 1912.[1] He could have added that whatever was *not* done at the Washoe smelter in attempting to abate sulphur dioxide and other emissions would also be highly beneficial to the industries.

If the aim of the stipulation—to ultimately eliminate deleterious fumes, particularly sulphur dioxide—was achieved at the Washoe smelter, every mineral smelter in the country would be obliged to underwrite the cost of achieving the same thing. The Washoe smelter at Anaconda then became the focal point of the industry for the development of air pollution abatement techniques or processes. By the time of Hammond's report, the commission had examined the Thiogen process, the Kelley process, and the Bradley process, as well as a number of others, but the board of experts' judgment on all the processes, Hammond said, was that, "as yet," they had not been "developed along practical lines to such an extent as to give assurances of being useful in remedying the situation complained of at the Washoe Smelter."[2]

Of the six experts engaged by the commission to assess the various processes, four were chemists and metallurgists of Anaconda Copper and two were physical chemists on "furlough" from the Bureau of Mines while working for the Westinghouse Electric Company. The Anaconda Company had extended complete cooperation and Hammond felt "decided progress" had been made in plans for collecting "solid dust" emissions. But in regard to "the hope of the ultimate elimination of sulphur dioxide," the inquiries and investigations of the board had not "pointed to any practicable way of eliminating these gases." The board thought, Hammond explained all too accurately, that "it seems not unlikely that some time may elapse before any thoroughly practical results may come from this work."[3]

After receiving this report in December 1912, the federal government withdrew from participation in the proceedings for almost a decade, except to the nominal extent the commission represented it. This allowed the mining and smelting industry the opportunity referred to earlier by the spokesman for the Bureau of Mines: Because the mineral industry was "one of the great basic industries of the country," it was "entitled to full consideration," and "should be accorded freedom to work out the smoke problem to the benefit of all concerned." Although withdrawing from active participation, the Justice Department nonetheless envisioned a time in which suit would be brought against Anaconda Copper for compensatory damages once the emission problem was remedied. Writing in 1914 to President Wilson's solicitor general, Ligon Johnson explained the role of the Stanford professors in the case and revealed the department's intentions:

"These gentlemen filed one copy of a joint report of their investigations which went to the Anaconda Commission for their use so long as necessary, the report to be returned for use by the Department of Justice when the time is ripe for action for damages. How soon such action will be filed must depend to a degree on the speed with which the Commission remedies fume conditions."[4]

What the government envisioned as a solution during the administrations of Roosevelt and Taft was an abatement of poisonous emissions—even though costly—through creative technology, with some help from the marketplace. They were not asking for a flat commitment to pure altruism but merely a recognition of some responsibility to society besides the single-minded purpose of maximization of corporate profits. They thought in terms of the long run as opposed to the short run. What the executives of Anaconda envisioned and what the members of the commission came to adopt as a solution was quite different.

As time separated the commission from the spirit of the 1911 stipulation, the commission's chief concern merged with that of the Anaconda Copper Mining Company.[5] Any means of abatement had to assure beyond question a beneficial economic result for Anaconda Copper. The company was unwilling to consider any cost in industrial efficiency to accomplish the ends of the stipulation. When the Justice Department withdrew involvement in the question, definition of the language of the stipulation fell to the commission: Terms such as "at all times complying," "promptly to comply," "well and truly comply," "best efforts to prevent, minimize and ultimately to completely eliminate," "practically applicable," "best calculated," and "a breach of the terms" became theirs to define.

As a result, the commission emphasized not what was *achievable*, but what it thought was *feasible*. The language of the stipulation was regularly employed; the connotations, however, changed. Consequently, the strenuous, expensive, and dedicated efforts of the Roosevelt and Taft men toward effective emissions control were reduced to exercises in futility, and the agreed-upon stipulation was twisted into a sham.

The commission's second report, written by a metallurgist in the Bureau of Mines, did not come for almost a decade. Even allowing for the effects of World War I, the length between the reports belied the commission's commitment to implement the terms of the stipulation. The report in fall 1920 was at best a lengthy and technical apology running 119 pages and explaining why the Anaconda Copper Mining Company had not done what it had covenanted to do. The 1920 report was replete with references to the "hearty and cordial cooperation" tendered the commission by the smelter's staff and sprinkled with allusions to the "best efforts" of the company.[6]

Numerous recommendations and suggestions were discussed in the report, invariably followed by rationalizations of why they could not be complied with. The persistent reason given was that the recommendation or process was not "commercially feasible" or "not yet commercially feasible." Yet no mention of commercial feasibility per se was contained in the stipulation. The closest the language of the covenant came to this idea was in calling for the defendant to conduct its operations and introduce methods and processes certified by a majority of the board "to conform to the most scientific processes then practically applicable to works of the character of those of the said defendant." This would seem to refer to inherent engineering or technical problems of the Washoe plant that would invalidate installation of a certain process.

The concept of commercial feasibility was introduced by Arthur E. Welles of the Bureau of Mines, author of the report. He reached back to 1917 and extracted the term from a speech given by Ligon Johnson before the New York section of the American Institute of Mining Engineers. At that time, Johnson had been "smoke counsel" for the American Smelting and Refining Company for about four years—at twice the salary he had been receiving from the federal government.[7] After quoting the stipulation in its entirety, Welles immediately followed by declaring the position of the Department of Justice at the time of its actions against the smelters. He quoted from a portion of Johnson's speech wherein he had remarked, "We proposed that the [Washoe] smelter operate under the best type of methods and appliances known to science and commercially feasible at the plants."[8] Thereafter, in the commission's deliberations, "commercial feasibility" became an iron law, the connotation of which was far from the "costly but necessary" view espoused by the Roosevelt and Taft men. Abatement processes that were not tied to efficient production were handily dismissed. Thus, the chief concern and aim of the commission was turned away, by Welles's sleight of hand, from the original objective of the stipulation.

Reporting on the "progress" of the commission up to 1916, Welles revealed that the flue and stack system "were in 1911 totally inadequate to condense and settle the finest of the dust and fume" and that the stack was "overloaded." The amount of arsenic emitted from the stack, according to the commission's figures, was over forty tons per day in 1911. Of all the arsenic treated at the plant in that year almost 87 percent of it blew out the top of the stack. Five years later, in 1916, over sixty-two tons of arsenic went through the stack daily and almost 89 percent of the arsenic that entered the plant left through the stack. In 1912, and presumably the same figures would hold for the years up to 1916, only the richest 25 percent of the arsenic was treated for the purpose of making fertilizer and other products, and 60 percent of that 25 percent found its way, along with a host of other emissions, out of the stack and over the town of Anaconda and the surrounding countryside.[9]

In regard to the sulphur dioxide problem, the commission recognized, Welles wrote, that because of the "enormous quantity involved," it was "by far the more difficult of the two [problems]" and "probably the most important from the standpoint of the community interests." Community interests were left undefined and never mentioned again in the report. "Ultimate success," Welles continued, "must depend not so much on working out new technical details at Anaconda for the recovery of the sulphur dioxide as upon opening up and reaching new and adequate markets for the products," regardless, apparently, of community interests.[10]

When the commissioners first started their work at the Anaconda plant, Welles said, "the average amount of sulphur being discharged per day from the stack was about 1,000 tons." The next year, operating under "normal output," 1,100 tons were discharged per day; by 1916, while "operating under maximum tonnage," the stack spewed out over 1,700 tons per day. In short, 70 percent more sulphur dioxide was being emitted than had been emitted five years earlier when the copper company signed the contract to "at all times use its best efforts" to eliminate sulphur dioxide.[11]

The commission had, however, investigated the Kelley process, the Hall process, and a number of others. None had proved attractive, the commission's spokesman explained: "Because of the necessary consumption of a comparatively large amount of fuel, which is relatively high priced at Anaconda, the reduction of sulphur dioxide at Anaconda does not appear attractive. It would seem poor conservation to attempt to force such a development on a large scale with its attendant high fuel consumption while there is any hope of getting a large part of the sulphur by a distillation method." The problems had been numerous, the report emphasized, but the commission would maintain its "lines of attack."[12]

The maintenance of the lines of the attack was small comfort to some of the farmers and ranchers in the valley who were trying to endure while the commission, with the assistance of the Bureau of Mines, investigated and experimented. W. R. Roberts, crop reporter for the Warm Springs District, wrote a letter to the secretary of agriculture in 1914 asking for federal assistance: "Send us a Government Expert," he asked, "to examine our cattle and horses, that is [sic] getting smoked to death by the fumes of the Washoe Smelter. Some of the horses' noses is [sic] almost eaten off by arsenical poisoning from the smoke." The state veterinarian had examined the horses and concurred, Roberts claimed. "Mr. Sec., some of us have been fighting the Washoe Smelter by Law for years, spent all the cash we had, and mortgaged our homes to see if some justice could be had. . . . Now the question is can we find some way to bring that murdering Company to justice."[13]

Referring to the inadequacy of the flue and stack system, which the commission had recognized three years earlier, Roberts described its effect on the valley: "When the wind blows down the valley you cannot see the valley for the smoke and since the [Bliss] decision they put in a new furnace several feet larger than any of the other furnaces. They open the big steel doors at the end of the flue wide open . . . and now . . . [the flue system has] a clear draft and we are certainly getting the contents." Derisively Roberts told of the efforts of Prof. F. G. Cottrell, who was assisting the commission and the smelter to solve the smoke problem:

> There is a gentleman here by the name of Cottrell to look into the smoke question, but he is not doing any good to relieve us of the smoke. He made a speech at Butte School of Mines 2 or 3 weeks ago that his patent is a success on the smoke problem. But he is going to figure out the loss of 10 percent in the tailings, and the smoke problem would come next. Him and the Company are putting in some very large buildings for the tailing process but nothing doing to save the poisonous substances coming out of that flue which is killing our cattle and horses; the tailing from the slime and 10 per cent copper loss is not doing us any injury.

On behalf of neighbors who shared his stubborn resistance to the course of the "smoke problem," Roberts asserted proudly: "We have done our very best, we do not want anything of the government. But we do ask of them to see if justice can be given us."[14]

Another rancher from Racetrack Creek, Montana, writing in a hand obviously unaccustomed to the pen, wrote the secretary of agriculture about how his stock was sick and "their noses all eaten out and full of ulcers." Asking the secretary's indulgence, J. T. Roseborough said,

> I don't know how hardly to address you but will try like this about the Washoe Smelter with all of those poisons we are breathing in the air causing sickness and death. . . . Remember we have taken up these lands for our homes and to make a living and to think of being run out and drove off of them without a penny and [know] not what. . . . If you think anything of your fellow citizenry and the farmer of the Deer Lodge valley you will take [immediate] steps to protect us at once.

Roseborough signed his letter, "Your democracy friend."[15]

The reply came from the Department of Justice, and it was prompt and crisply officious: "The Government is conducting investigations with a view to ascertaining means to abate the poisonous fumes and injurious conditions of which you complain. . . . Aside from what it is doing in behalf of its own public interests and the ultimate effect thereof on conditions affecting your private interests, the Government can take no action in your behalf."[16]

The farmers in California meanwhile, notwithstanding the agreements reached with the smelters, themselves, and the federal government, had fared little better according to the leader of the Shasta County Farmers' Protective Association. Writing to the attorney general in late 1912, the head of the farmers' association claimed that the baghouses installed at the Mammoth smelter were "as practical for controlling sulphur fumes as a sieve, and the damage is going merrily on." The farmers insisted they were not "fighting mines nor miners, neither do we want to do any injustice to any smelting interests." "But only want the damaging effects eliminated, and we shall be entirely satisfied," Chub Gibson wrote on behalf of the farmers. The government, in return, explained that its experts had informed them that the smelters were not damaging the national forests and their authority and interest in the matter stopped at those boundaries.[17]

The California farmers persisted, however, and eventually the Department of Justice provided them with a rationale that is of interest because it expresses the dilemma of the reformers and the limits of the law as then constituted. The department thought it unfortunate that the United States did not have the necessary authority to intervene in an area of "fume influence" where it was not the owner of the property or where the smelters were not willing to submit to inspection and supervision. The government's powers were constitutionally limited in any conflict between it and private industry and in conflicts within the private sector. The government could not be the partisan of either the smelter or the farmer anymore, it explained, "than it could lend its aid to one farmer in a proceeding against another farmer."[18]

There were certain hard realities that could not be obviated in a contest with the smelters by groups of individual citizens. In such litigation, the Justice Department said as tactfully as possible, rarely do the plaintiff and defendant meet under equal financial conditions. Drawing on its long, frustrating experience, the department commented on the role of experts and their influence on the court: Experts were essential in fume suits because of the complex technology and scientific questions involved. And the court was forced to be guided by the experts' testimony, which testimony, the writer emphasized, in addition to being expensive, confusing, and complex was "sharply conflicting." "Erroneous statements of experts," he said cautiously, "and lack of technical knowledge on the part of trial judges, sometimes does result in a mis-carriage of justice."[19]

Nevertheless, the Department of Justice knew of no other way, the spokesman concluded, to avoid "this heavy expense and danger of misleading statements" than to turn the problems over to "proceedings similar to those adopted by the government in the suit of the United States *vs.* The Anaconda & Washoe Copper Companies." In this way the complexities of the problem were removed from the court and submitted to a commission of experts composed of the director of the Bureau of Mines and "two other men of the highest character and technical attainment."[20]

Subsequently, the governor of California appointed a commission consisting of the secretary of the State Board of Health, the horticultural commissioner, and the state veterinarian to investigate the problem of smelter emissions in Shasta County.[21] And the controversy dragged on.

The extensive refining, marketing, and merchandising pattern created by Anaconda Copper in the twenties suggests that the problem in California was ultimately solved by shipping the sulphur ores to Yerington, Nevada, and from there shipping the copper precipitates to Anaconda's smelter in Montana.[22]

By 1917 the Anaconda smoke commission had decided on a plan to improve the recovery system, ostensibly to reduce arsenic emissions. Arthur Welles explained the plan boldly: All the gases would be put "through a single large installation" of Cottrell electrical precipitation units, rather than through several installations scattered about the plant. With the "final plan . . . definitely decided upon," the company submitted a request for appropriations to its New York office. In addition to the Cottrell units, the plan called for a new stack reaching 525 feet into the sky and a modification of the flue system to accommodate the stack and the Cottrell units. "The total original estimate for this installation was $1,600,000," Welles reported proudly, and the appropriation was "immediately obtained." In addition, Welles pointed out, the brick plant was expanded, increasing its capacity and necessitating an additional cost of ninety thousand dollars. Construction work began in mid-1917, and the stack was completed in late 1918. Problems, however, were encountered in completing the Cottrell system that was intended to allow the smelter to expand its phosphate program and reduce the sulphur emissions.[23]

In late 1918, Cornelius F. Kelley, then vice president of Anaconda, wrote a long letter to the director of the Bureau of Mines in which he put forth a variety of reasons why the company had not complied with the 1911 stipulation. "Inasmuch as the Anaconda Copper Mining Company was delayed carrying into execution the experiment recommended by the Smoke Commission, for the extension of the phosphate program, at the Washoe Reduction Works," Kelley admitted with uncharacteristic candor, "I desire to submit the following as explanatory of such delay." There followed a listing of numerous adverse circumstances imposed upon the company as a result of the world war and the demands of national interest. Anaconda had been forced to work under "serious difficulties" as a result of "the shortage of sufficient men to carry on the necessary work of producing as much copper and high grade zinc as possible for governmental purposes," Kelley explained. In spite of "the high wages that are being paid" for common labor, "the labor shortage in Montana has reached a rather critical stage." To compound the difficulty, there had been "a marked decrease in efficiency" of the current labor force. In addition, "the technical staff has been greatly depleted, many of our most valuable technicians having enlisted in different branches of the National service." This left the company "with scarcely a sufficient number of trained men to carry on the routine requirements of the business," Kelley explained.[24]

Consequently, Kelley continued, because of the excessive cost of labor and the impossibility of securing existing materials, "the cost of experimentation" had become "unduly expensive," thus delaying the installation of the Cottrell system. He then explained that "the present unsettled economic conditions" made it infeasible to engage in the phosphate business: "We believe that it would not be practical," Kelley said, "to commence the construction of a large phosphate plant under present conditions."

There were a number of other adversities, too, but these were the most compelling, according to Kelley. Nevertheless, the executive assured the director of the Bureau of Mines, "If the Commission should, notwithstanding these conditions, believe that we should proceed at once with the experimental plant heretofore recommend, we will cheerfully comply with your recommendation, and use our utmost endeavor to carry into effect as rapidly as circumstances will permit."[25]

As it turned out, the commission did not believe they should. It was too risky. As late as 1924 the director of the Bureau of Mines was incorporating Kelley's comments into his report to the U.S. attorney general on the progress of the commission.[26]

Not mentioned by vice president Kelley in his letter was the financial position of the mining company during this period of "hardship." Between 1915 and 1918, when Kelley wrote his letter to the director of the Bureau of Mines, the company had reaped a cornucopia of war profits. Within those three years, Anaconda's coffers were filled with $122 million in profit—an amount greater than the company's total assets were the year before it contracted with the U.S. government to abate air pollution. In 1918, the company's assets were double those of 1910, amounting to $237 million, and the company's surplus had multiplied twenty-two times, leaping from $3 million to $66 million. Dividends paid for the four-year period between 1913 and 1917 totaled $48,884,375; roughly 36 percent of this total had been disbursed to shareholders the year prior to Kelley's hardship letter to the director of mines.[27]

Furthermore, Anaconda executives were not looking at an end to the prosperity. They were planning for as great a postwar boom in copper demand as had taken place during the war. President John D. Ryan had prophesied two years earlier, in 1916, that "the resumption of peace in

Europe will, with perhaps a brief intervening period of readjustment, be accompanied by an enormous demand for copper." Preparing to take advantage of the anticipated boom, Anaconda Copper was channeling billions deep into the Atacama Desert in Chile through the Andes Copper Company, which functioned as a holding company. Control of promising mining properties had been acquired and construction had begun on a fifty-five-mile railway and a thirty-five-mile pipeline to tap what one writer described as the stuff of an engineer's "dream of the ideal mining property."[28]

The company's financial situation was further enhanced by virtue of the most favorable tax rate of any mining state in the West. Since 1916, to cite just one example, the mines of Arizona had been assessed roughly three times as heavily as had the Montana mines.[29] Yet in the face of this rosy financial condition, Anaconda Copper maintained it was "unduly expensive," impractical, infeasible, indeed, "impossible" to invest an adequate portion of their wealth into means of abating the morbid effluents of the Washoe smelter.

During the same period—1916 to 1918—while consumption, production, and surplus grew, emissions proliferated. "The amount of dust being discharged from the stack," the commission casually observed, "was much greater than in earlier years." Production of copper alone had almost doubled, jumping from around 13 million tons per month to 25 million during the three years. "With the increased volume of the gas and the increased temperature of gases in the flues there was, of course," the commission commented, "less tendency for the dust and fume particles to settle and a greater proportion therefore escaped through the stack into the atmosphere."[30]

There was bad news on another front, as well. New technology that provided more efficiency in the smelting process contributed to the increase in emissions, said the authorities: "The introduction of a new method of concentration, that is, the introduction of flotation methods, also brought about a greater amount of dusting in the roasting and reverberatory furnace and as the dust from flotation concentrates is extremely finely divided and in many ways resembles a fume, only a small portion of this settled out in the high velocities in the fume chambers and therefore the greatest proportion was discharged into the atmosphere."[31]

Figures comparable to those of the years 1911 through 1916 were not provided for the period 1918 to 1920, but the reports confirmed that sulphur dioxide emissions never did get below seventeen hundred tons per day. Illuminating, nevertheless, was the recovery Arthur Welles said the new system would effect. It was shades of the 1903 remodeling of the flue-and-stack system that was said to have been done to meet the complaints of the Deer Lodge "smoke farmers." About 80 percent of the copper, gold, and silver would be removed from the smoke stream and 60 percent of the lead; but the "estimated arsenic recovery would only be about 45 percent."[32]

Welles indicated the commission was sorry about the low arsenic recovery, but it was unavoidable: "It was the general opinion, both of the representatives of the Commission and the representatives of the company, that a greater Cottrell treater capacity would ultimately be required in order to get a satisfactorily high arsenic recovery. However, as the construction program then authorized, with the regular upkeep of the plant, would tax to the limit the construction capacity of the plant, the recommendations for the erection of additional treaters was not made at that time."[33]

Little was done to remedy the sulphur dioxide problem for the reasons provided in Kelley's letter: "It could readily be carried out on a large operating scale, whenever the project looked commercially feasible," the commission said. In concluding the report, Welles felt compelled to assure anyone who may have had cause to read it that "In all these investigations, whether aimed at the recovery of solids or the development of uses for the waste sulphur dioxide, the Company has afforded the most cordial assistance and cooperation to the Commission, and any recommendations or suggestions of the Commission have been carried out or followed up in a very satisfactory and vigorous manner."[34]

The last report was written in 1924 by the director of the Bureau of Mines, H. Foster Bain. It was sixteen pages long, most of it devoted to reviewing the previous report. Since 1916 the commission had had no staff at the Washoe. All the figures for their analyses had been supplied by the company. On the basis of these figures, the commission reported that for a three-month period in 1924, while the smelter was operating at 80 percent of capacity, it recovered through the use of the Cottrell precipitation apparatus 88.0 percent of the dust and fumes, 93.5 percent of the copper, 81 percent of the arsenic. Lead and zinc contents of the dust and fumes were not considered because the company claimed they were "largely wasted in the slag produced by the smelting operations." Besides, the report explained, "There is no revenue from these metals and none is likely under present conditions as the cost of recovery probably would be too high to warrant special processes to recover them." The report considered percentages of various solids recovered but not of the tonnage escaping through the stack.[35]

The marked increase in the recovery of arsenic was due in large part to the efforts of an industrious little gray beetle far removed from the Anaconda region. Independent of the efforts of the commission and of company officials, arsenic treatment had become commercially feasible, offering attractive returns and a promising future. As the commission explained:

> With the present demand for calcium arsenic for insecticide for fighting the cotton bollweevil and the constantly increasing use of arsenical materials for other purposes, arsenic has become, at least

> for the time, a valuable commodity. The value of white arsenic during 1923 has approached that of copper.
>
> Hence, further consideration has been given to the probable cost of obtaining a greater recovery of the arsenic from the slag and the dust and fume.[36]

While providing no figures, the report asserted that the arsenic that was escaping through the stack was less than a third of what it had been in 1916. This, it declared, "probably could be discharged safely into the atmosphere without endangering any outside community interest." The commission based its conclusion on the assertion that "No conditions have arisen since then which would indicate that this arsenic output could constitute a nuisance to the outside surrounding community." Hastily it added that, since the community interest had been safeguarded, "the Commission is of the opinion that all the danger of injury to outside interests by solid emanations has been removed." Further investment to reduce the arsenic emissions could not be justified, the commission claimed, because their studies had pointed to "so many uncertainties" in the market that "it would hardly be justified from an economic or practicable standpoint."[37]

Likewise, according to H. Foster Bain of the Bureau of Mines and the other members of the commission, because of an improving market picture, the future held promise for decreasing the emission of sulphur dioxide. But rather than employing the terms of the original stipulation, which called for the ultimate elimination of the fume, the report talked about the future "utilization" of sulphur dioxide to produce a high-grade fertilizer. The amount of sulphur dioxide being discharged from the smelter was not revealed. Instead, the report discussed the poor market that had existed before the war and had prevented Anaconda Copper from engaging in the fertilizer business. Even since 1920, "the fertilizer business of the country has been in a very unsettled and unsatisfactory condition," the commission explained, yet "During the last four years the Anaconda Company has prosecuted as vigorously as feasible the program for the production of the high grade superphosphate fertilizer, which was started in 1920, following recommendations by the commission, as a step toward going as far as practicable in the elimination of deleterious substances."[38]

As the farmer became better educated in the use of fertilizer, the commissioners observed, in time a satisfactory market "may be safely anticipated."

"Having $2,500,000 already invested in phosphate mines and plants, the Anaconda Copper Mining Company, without question, will develop the phosphate program as rapidly as possible without any further recommendation of the Commission."[39]

On that notice, the commission washed its hands of the air pollution problem at Anaconda Copper's Washoe smelter and indirectly of that of the entire smelting industry.

The land and livestock within the zone of smoke influence from the Anaconda smelter continued to suffer from the effects of arsenic emissions as late as the mid-1920s, as witnessed by the Anaconda Smoke Commission's report of 1924 and the complaints of farmers. Yet the literature on the morbidity of smelter emissions paints a different picture. As late as the 1950s, the literature was characterized by a convenient if not self-serving reduction that concealed the harsh realities. In discussing the injury to animals as a result of the arsenic emissions of the Anaconda smelter early in the present century, the *Air Pollution Handbook,* published in 1956, commented on the elimination of the problem: "Successful solution of the problem was finally achieved in 1910 by the installation of an enormous Cottrell electrical-precipitation system, and injury to livestock disappeared for all time," it claimed.[40]

Judging from the activities of Anaconda Copper in the two decades following the signing of the 1911 stipulation, despite the rhetoric of pious intent, the company had little if any intention of engaging seriously in investigating and developing measures to abate air pollution at their Washoe smelter. Unlike the commission and the federal government, its authority in the matter was plenary. It could exercise the good faith called for in the stipulation or it could not. Regrettably, it chose the latter course.

As the commission had seized upon commercial feasibility to explain away the lack of effective abatement processes, the executives of Anaconda Copper seized upon part of the language in a single clause and resolutely turned their backs on the rest of the contract. The stipulation called for them to operate their smelter in a manner best calculated to avoid the emission "of fumes containing deleterious matter injurious to the interests of the complainant"—the United States. If the federal government's prime concern, the well-being of the national forest, could be removed from the controversy, then the question of abatement would become moot. Accordingly, the copper company sought to remove the source of complaint against its smelter rather than to remove the fumes that provided the basis for the conflict. Doing so required a coarse indifference to the property and health of those inhabiting the area subject to fume influence, but these were of low priority in the reckonings of the men making the decisions for the smelting plant and overall interests of the Amalgamated trust.[41]

In like manner, the company moved to resolve the problem with the protesting farmers of the Deer Lodge valley. They bought out the agricultural areas deemed susceptible to smoke damage. Acquisition of the land posed little difficulty for the Amalgamated after 1910. Much of the land was owned by the Northern Pacific Railway as a result of the Railroads Subsidies Act of July 2, 1864.[42] But even more significant was the fact that as a result of appealing the decision of Judge Hunt in the Bliss case, the farmers were broke, dispirited, and vulnerable.

In mid-1910, about a year after the Bliss decision, the chairman of the farmers' association revealed their dilemma to the Justice Department:

"The Farmers' Association, I am sorry to say," Walter Staton wrote, "at this time have no funds in their treasury, in fact the expenses of perfecting the appeal has practically drained them of every dollar they possess, except in one or two instances. I, myself, have given a Trust Deed to my property in order to obtain funds to pay my proportion of the appeal in this case."[43]

A year later after the adverse decision in the appeals court at San Francisco, the association's counsel petitioned the U.S. Supreme Court for a writ of certiorari, a proceeding in which the higher court would review the lower court's procedure and decision.[44] In order for the record of the case to be printed and worked up for consideration by the Supreme Court, however, the hard-pressed farmers would have to post $4,155 to cover the cost of printing. The farmers were unable to raise the money and when the high court reached the case on the docket, it was dismissed because the necessary papers were not available.

Sometime before this dismissal, however, most of the farmers had recognized the hopelessness of their cause and had capitulated—or "sold out" in the minds of some of the more recalcitrant farmers. After the wealthier members and leaders of the association settled, the others followed suit. By 1924 the company owned the greater part of the area around the smelter assumed to be within the smoke zone and held the so-called smoke rights on practically all the farms in the Deer Lodge valley. "Smoke rights" in all likelihood operated as would an easement, allowing the right-of-way over another's lands or, like a negative easement, preventing a landowner from building a structure that causes the obstruction of light to a house on an adjoining property. Under smoke rights, presumably, the owner or lessee of land in the valley would be barred from pursuing an action at law against the Washoe smelter for damage as a result of smoke.[45]

It was the existence of this condition within the Anaconda area that allowed the director of the Bureau of Mines to report in 1924 with but a modicum of probity that "all the danger of injury to outside interests" had been removed, and at the same time allowed him to move to the conclusion that "so far as the safeguarding of the community interests is concerned, the Anaconda Company has accomplished much more than what was thought to be necessary."[46]

In view of the Anaconda's attitude, one may wonder why the company had bothered to sign the stipulation at all. The reasons were several and interconnected. The policymakers for the trust understood that the commission's role would be sterile. A commission controlled by men, regardless of the quality of their character, cannot realistically be expected to call to account and invite government intervention in an enterprise in which they had established careers and heavy investments. Amalgamated executives had also not missed the significance of the failure of the Interstate Commerce Commission to win suits against the railroads during the years 1885 to 1905.[47] In addition, the trust counted on the suasion of "the magnitude of interest" to detour the government from instituting proceedings that would lead to an injunction that in turn would cause adverse economic and political repercussions. But most importantly, signing the stipulation was the most expedient way for the company to remove the government from involvement in their industry and to keep regulatory activity at bay.

The postwar years afforded a most propitious time to terminate remaining federal interest in the conduct of the Washoe smelter. In the wake of the war, the progressive concerns of the Roosevelt, Taft, and Wilson administrations were subverted to the promise of a new era of enterprise and the vision of the good life. Emerging from high positions within Wilson's administration during the war, the nation's corporate executives went back to their firms with an idea singing in their heads. The war had taught them that federal involvement need not be antibusiness. The president of the Anaconda Copper Mining Company,

John D. Ryan, himself served as assistant secretary of war and worked closely with Bernard Baruch, chairman of the War Industries Board.[48]

The new thinking that pervaded the twenties was perhaps best expressed by Herbert Hoover, who presided over the Department of Commerce for most of the decade—from 1921 to 1929. When he became president, Hoover promoted the idea of "associational activities" to eliminate inefficiency and waste that accrued from rigorous competition. Under this concept, the government would actively cooperate as a partner with the major corporations and groups of firms drawn together in trade associations. It was to be a "new era of national action, in which the federal government [would form] an alliance with the great trade associations and the powerful corporations," Hoover said. As he saw it, the country's economy "was passing from a period of extreme individualistic into a period of associational activities."

Under federal aegis and industrial efficiency, and abetted by the hucksterism of a burgeoning advertising business, production rose and the standard of living for most Americans climbed to new heights. Until the latter months of 1929, it was a good time to be in business.[49] It was also a time peculiarly congenial to the plans of Anaconda Copper to end everything that remained of the federal government's involvement in their Washoe smelter and consequently in the industry in general.

In the cooperative spirit of the new era, certain associational activities took place in 1923 between the mining company and the U.S. Forest Service in Montana. The association, however, was one in which the government functioned as the handmaiden of a powerful corporation rather than as its partner. The opportunity to erase the contretemps and commitment of 1911 and provide for the future had presented itself to the company.

A series of conferences was initiated in the fall of 1923. According to Vice President James R. Hobbins of the Anaconda Copper Mining

Company, the conferences took place upon the suggestion of the district forester in Missoula. The purpose of the conferences, the executive explained in a letter to the head of Forest District 1, was to "consider the advisability of disposing, by exchange of lands . . . any and all claims which might be urged by the U.S. Government for damages claimed to have been sustained to lands of the Forestry Department through the smelting operations of the Anaconda Copper Mining Company near Anaconda, Montana." Also discussed at the conferences, Hobbins said, was "the prevention of future controversies arising through the operations of such smelting plant, by the exchange of lands in the vicinity of said smelting, for timber lands in Western Montana."[50]

At the conferences, the executive explained, it had been agreed that this proposition "be formally instituted by means of a letter from the Anaconda Copper Mining Company covering, so far as possible at this time, the matter of the proposed exchange." Accordingly, the vice president went on to describe the details of the exchange and how it should be executed. One thing the company wanted clearly understood, he told the forester, was the basic principle upon which the transaction was to rest: "It is understood that the basis of the exchange shall be that each party shall convey equal value, such value to be determined by agreement and agreed appraisals by the parties of the land and timber thereon."

In addition, the company wanted it clearly understood just exactly what it expected in return for having participated in the exchange: "As stated to you in Missoula," the district forester was informed, "with the completion of this transfer we would expect a release and discharge on the part of the Government, releasing the Anaconda Copper Mining Company from all claims arising prior to the date of the settlement, because of injury to land, timber, etc. from the operation of the smelting and reduction plants."

No problems were presumed to exist, according to the Anaconda official, except the apparent minor one of providing for a federal law that would legitimize the process and principle. "It is understood," he repeated,

> that the carrying out of such exchange will probably necessitate passage by the federal Congress of additional legislation conferring necessary authority upon the Forestry Service and other proper governmental officials. Such legislation should not alone confer full authority upon the proper governmental officials to convey the land proposed to be conveyed . . . but should also give the governmental officials authority to hereafter compensate by conveyance of other lands for such loss as may be sustained by the patenting of mining claims . . . or for any other loss of acreage which may be attempted to be conveyed by it, which the Anaconda Company may not . . . have acquired title to, or may hereafter be deprived of.[51]

Obviously, the Anaconda Copper Mining Company was seeking much more than the termination of legal responsibility for the damage done from about 1900 to 1923. Unabashedly it sought sovereign authority for the right not only to pollute the air in the Anaconda region but for the right to pollute or damage the public domain whenever, wherever, and however it chose. Apart from the question of wise public policy, considering that as of 1916 the Anaconda Company owned almost $5.5 million worth of timberland that provided a veritable reservoir for endless trades, the idea of codifying "exchanges" by statute would amount to issuing the company an unrestricted license for pollution.

While the discussion and conferences were taking place in Montana, similar activity was occurring in Washington, D.C. With Attorney General H. M. Daugherty opening doors, legal counsel for the copper company was meeting with various administrative heads and lesser officials to argue the Anaconda case. Several days after the Anaconda executive made his proposal to Fred Morrell, the district forester in Missoula,

William B. Greeley, the Forester in Washington, D.C., wrote Morrell requesting information on the investigations carried out by the Roosevelt and Taft administrations and asking for his views on the idea of a "proposed exchange" with Anaconda Copper. In the meantime, the powerful liberal senator from Montana, Thomas J. Walsh, introduced Senate Bill 582, calling for government authority to exchange privately owned lands adjacent to national forests in Montana.[52]

Under the terms of Senator Walsh's bill, the National Forest Consolidation Law of March 20, 1922, would be expanded to authorize the government to exchange national forest lands for private land of equal value situated either within a national forest or within six miles of the boundary of a national forest in the state of Montana.[53]

To what extent Senator Walsh and the department heads, Secretary of Interior Hubert Work and Secretary of Agriculture Henry C. Wallace, appreciated the full import of the bill in relation to the aims of the Anaconda Copper Mining Company is not known. Henry Wallace's stated reason for approving the bill was that it would serve to protect the national forests and improve the administration of the forests, thereby benefiting the public interest. Referring to private lands within and near national forests, he wrote the chairman of the Committee on Public Lands and Surveys, of which Senator Walsh was a member, that "Quite frequently in their unprotected state they present a menace to the lands owned by the Government and increase the cost of protecting the Government lands from fire. The consolidation of these areas under Government administration would materially reduce the danger from fire and would simplify many problems of administration." Secretary of Interior Work, claiming there was a "great demand for homestead lands, and the area of public land in Montana [was] being rapidly diminished," saw the bill as an opportunity for removing grazing lands from the claims of homesteaders. He amended the bill accordingly and recommended its passage. Senator Walsh's motives for introducing the bill cannot be known.[54]

The bill was directed specifically at private land situated within six miles of a national forest boundary and, significantly, affecting only such lands in Montana. Apparently, the lands designated for the trade by Anaconda Copper lay outside but near national forests. If the bill was advantageous to the administration of the almost 16 million acres of national forest land in Montana, why was it not as equally advantageous to the administration of over 158 million acres of national forest lands outside of Montana? The silence of high-placed government officials on this question points, if not to malfeasance, then to nonfeasance. And the specter looms of prominent officials engaging in convenient and studied indifference to a fundamental principle of public policy in order to accommodate the specific interest of one ruthlessly powerful corporation, for at approximately the same time, certain members of Congress were endeavoring to further the principle of forest conservation in the form of the Clarke-McNary Law, hailed as a milestone by Greeley in 1924.[55]

About a month after Senator Walsh introduced his bill, and weeks before the Agriculture and Interior Departments supported it, the district forester, in Missoula, Fred Morrell, sent his report to Greeley in Washington. It was a peculiarly constructed report. Although not directly opposing the Anaconda proposal, it was not an unqualified or enthusiastic endorsement of the plan. Framed in a cleverly ambiguous fashion, it left room for a decision to be made either for or against. Morrell gave a summary of each of the investigations carried out by the earlier administrations; he cited their conclusions that the sulphur oxides from the smelter were responsible for the damage and reported that the last investigations had demonstrated that the damage was increasing. He admitted that "reliable figures on present damage" were difficult to come by but that an "enormous quantity of sulphur" was being emitted. If the smelter increased its output "very materially," Morrell observed, there would be a recurrence of the "old damage" caused by arsenic. Morrell characterized the proposed exchange bluntly: "The turning of all Forest Lands which have been damaged to the A.C.M. Company would be really making the same kind of deal as has been made between the A.C.M. Company and the owners of farm lands," he wrote.[56]

Within weeks after receiving the report of the district forester, Greeley began speaking of the Anaconda case as a "trespass problem" and prepared to make the exchange. The next month the acting secretary of agriculture requested a legal opinion from Attorney General Marvin F. Stone, asking what effect the exchange would have on the pending suit filed by the United States in 1910. "The Forest Service," the secretary wrote Stone, "looks with favor upon this plan and believes that it offers the best solution of the problem." No mention was made of the stipulation of 1911, only that a commission of well-known mining experts and the director of the Bureau of Mines had been "appointed to investigate the matter."[57]

"It would seem to me," Attorney General Stone replied two months later,

> that the consummation of the exchange suggested would practically eliminate the basis upon which the suit is predicated, for the damage to the timber and to the forest cover over the lands within the national forests furnished the grounds for the suit. There was some allegation in the bill that one of the results of the emission of the smelter fumes was that which followed the destruction of the forest, namely, the unusual discharge of flood waters because of the destruction of the forest cover and the consequent injury to the headwater of navigable streams. . . . I assume that the proposition which was made by the defendant companies is for the exchange of lands of a value commensurate with the value of the government's lands before they were damaged. This being the case, I can see no objection to the settlement of the case on the lines suggested.

There were, however, certain things to guard against, Stone said. "It might be well, however, to guard carefully any agreement for settlement by appropriate stipulation that it is to cover only the claims for damage to the lands and timber of the United States thus not making it an unconditional release for all claims and demands arising out of the maintenance of this nuisance by the defendants, unless your investigations have satisfied you that no other substantial damage to the United States has accrued."[58]

The department was satisfied not only that no more damage had accrued but that none would accrue. The bill slipped through Congress without any perceptible resistance, becoming law on February 28, 1925.[59] Consequently, any remaining incentive for the Washoe smelter or for any smelter in the industry to abate air pollution was removed. No longer was control of smelter emissions a part of conservation policy. The federal government in the twenties, under the policy of associational activities, achieved, not the symbiotic relationship envisioned by Hoover, but one more akin to parasitism in which the great corporations claimed all the advantages and rights but practiced none of the responsibilities. It returned to the conduct of the smelting industry in the nineteenth century. The pattern was set: increased production, greater pollution, and lucrative real estate deals.

Work on the exchanges went forward immediately after Senator Walsh's bill became law. Much of the investigative work and appraising of the lands had been done prior to passage of the law, some as early as 1921. A total of six separate exchanges took place. All but one were consummated between 1928 and 1932. Title was passed on the last one in 1937. When the negotiations were completed and ownership finally transferred, the result conformed to the idea expressed by Cornelius Kelley twenty-seven years earlier rather than to the estimate Ligon Johnson made in 1914.

Johnson had negotiated the McCune timber depredation suit against the Anaconda Copper Mining Company and had settled it for $150,000 in 1914. He said at the time that the area involved was but "a small part of that involved under the Government's claim for fume damage." He cited it in an effort to convey to the Department of Justice some concept of the total amount of money involved in the final adjudication. One might then have reasonably envisioned a figure of $1 million to $2 million, perhaps more. Kelley's view was that "the government should be content to accept a minimum sum" in final settlement of the entire case. He proposed a $25,000 lump sum plus $5,000 a year for additional damages. By the time of final settlement, this would have totaled $160,000. Assuming that the value of the dollar in its capacity to purchase forestland had slipped roughly 50 percent since Kelley made the offer, the $160,000 in 1937 would amount to $320,000.[60]

Attorney General Wickersham had dismissed Kelley's offer at the time as "absurd." The government of the twenties, however, judging from the amount finally negotiated, would have considered the offer magnanimous. For the total value of lands selected in settlement of the case the government was content to accept $227,311. The Justice Department was not involved in the exchange and no release of all claims was ever executed.[61]

In classic symmetry, the denouement of the struggle to compel copper smelters to abate poisonous emissions ended where it had begun. In 1933, within the gloom of the fortresslike courthouse at Butte, Montana, the clerk of the federal district court wrote the final entry in the docket opposite the case of the *United States of America v. Anaconda Copper Mining Company*: "Abandoned."[62]

– Notes –

Introduction

1. K. Ross Toole, *Twentieth Century Montana: A State of Extremes* (Norman, Okla., 1972), 289-90. See also K. Ross Toole, *Montana: An Uncommon Land* (Norman, Okla., 1959); Joseph Kinsey Howard, *Montana: High, Wide, and Handsome* (New Haven, Conn., 1943). For a perceptive view of Toole the man and the historian, see Harry W. Fritz, "An Uncommon Man, K. Ross Toole, 1920–1981," in *Montana and the West: Essays in Honor of K. Ross Toole*, ed. Rex Myers and Harry W. Fritz (Boulder, Colo., 1984), 4-16.

2. Michael P. Malone, Richard B. Roeder, and William L. Lang, *Montana: A History of Two Centuries* (Seattle, 1991), 378-82. On water preservation, see William L. Lang, "Saving the Yellowstone," *Montana The Magazine of Western History*, 35 (Autumn 1985): 87-89.

3. On the effects of the coal issue on Montana politics, see K. Ross Toole, *Rape of the Great Plains* (Boston, 1976), 208-26; James Conaway, "The Last of the West: Hell, Strip It!," *Atlantic Monthly*, September 1973, 91-103; Louis D. Hayes, "Who Will Control Montana's Coal?" *Montana Business Quarterly*, 13 (Spring 1975): 26-32; *The Plains Truth*, the Billings, Mont., Northern Plains Resource Council monthly publication (1972–76); Richard L. Reese, ed., *Coal Forum* (Missoula, Mont., 1974); Gray Wicks, "A Look at Coal-Related Legislation Enacted by the Forty-third Legislative Assembly," *Montana Business Quarterly*, 11 (Summer 1973). On the national trend toward use of western coal during the 1970s, see Bruce Ackerman and William Hassler, *Clean Coal/Dirty Air* (New Haven, Conn., 1981).

4. On the grass-roots efforts and an admiring interpretation of the 1973 legislative session, see Toole, *Rape of the Great Plains*, 216-21.

5. On the history of electrostatic precipitation and its relationship to smelters, see Myron Robinson, "Electrostatic Precipitation," in *Air Pollution Control Part I*, ed. Werner Strauss (New York, 1971), 228-34. For a superb discussion of the importance of new legal standards in the early twentieth century and their application in the Anaconda smoke cases, see Gordon Morris Baaken, "Was There Arsenic in the Air? Anaconda Versus the Farmers of Deer Lodge Valley," *Montana The Magazine of Western History*, 41 (Summer 1991): 30-41.

6. For the history of smoke abatement and early U.S. antismoke legislation, see W. F. M. Goss, *Smoke Abatement and Electrification of Railway Terminals in Chicago* (Chicago, 1915), 28-31; R. Dale Grinder, "The Battle for Clean Air: The Smoke Problem in Post–Civil War America," in *Pollution and Reform in American Cities, 1870–1930*, ed. Martin Melosi (Austin, Tex., 1980), 83-101. For historic efforts in England, see Carlos Flick, "The Movement for Smoke Abatement in Nineteenth-Century Britain," *Technology and*

Culture, 21 (January 1980): 29-50.

7. Paul U. Kellogg, ed., *The Pittsburgh District: Civic Frontage* (New York, 1914); John O'Connor, Jr., "The History of the Smoke Nuisance and of Smoke Abatement in Pittsburgh," *Industrial World* (March 24, 1913). On the Pittsburgh Survey, see Joel A. Tarr, *The Search for the Ultimate Sink: Urban Pollution in Historical Perspective* (Akron, Ohio, 1997), 77-102.

8. Goss, *Smoke Abatement*, 38; Robinson, "Electrostatic Precipitation," 233-39. On technology and the Progressive Era conservationists, see Samuel P. Hays, *Conservation and the Gospel of Efficiency: The Progressive Conservation Movement* (Cambridge, Mass., 1959), and for a recent reinterpretation of Hays's thesis, see Clayton R. Koppes, "Efficiency, Equity, Esthetics: Shifting Themes in American Conservation," *Environmental Review*, 11 (Summer 1987): 127-46.

9. Samuel P. Hays, *Explorations in Environmental History* (Pittsburgh, 1998), 222-47. On the electron capture device, its inventor, and its importance, see Lawrence E. Joseph, *Gaia: The Growth of an Idea* (New York, 1990), 22-23, 156-58.

10. Hays, *Explorations*, 224-28. For principal content and specific chemical standards of federal air quality legislation, see Kenneth Wark and Cecil F. Warner, *Air Pollution: Its Origin and Control* (New York, 1981), 42-52. For an example of the relationships between politics and enforcement of the Clean Air acts since 1967, see Charles O. Jones's Pennsylvania case study, *Clean Air: The Policies and Politics of Pollution Control* (Pittsburgh, 1975).

11. For evaluations of environmental history and bibliography, see William L. Lang, "Using and Abusing Abundance: The Western Resource Economy and the Environment," in *Historians and the American West*, ed. Michael P. Malone (Lincoln, Nebr., 1983), 270-99; Richard White, "American Environmental History: The Development of a New Field," *Pacific Historical Review*, 54 (Summer 1985): 297-335; Samuel P. Hays, "From Conservation to Environment: Environmental Politics Since World War II," *Environmental Review*, 6 (Winter 1982): 14-41.

12. Martin Melosi, *Garbage in the Cities: Refuse, Reform, and the Environment, 1880–1980* (Chicago, 1981); Tarr, *Search for the Ultimate Sink*; Craig Colten, *Industrial Wastes in the Calumet Areas, 1869–1970* (Urbana, Ill., 1985).

13. Joel A. Tarr, "Changing Fuel-Use Behavior and Energy Transitions: The Pittsburgh Smoke Control Movement, 1940–1950," in Tarr, *Search for the Ultimate Sink*, 227-61; Joel Tarr and Carl Zimring, "The Struggle for Smoke Control in St. Louis: Achievement and Emulation," in *Common Fields: An Environmental History of St. Louis*, ed. Andrew Hurley (St. Louis, 1997), 199-220; Samuel P. Hays, *Beauty, Health and Permanence: Environmental Politics in the United States, 1955–1985* (Cambridge, England, 1987). On the Donora, Pennsylvania, disaster, see Mark Neuzil and William Kovarik, "The Importance of Dramatic Events: The Donora Killer Smog and Smoke Abatement Campaigns," in *Mass Media and Environmental Conflict: America's Green Crusades* (Chicago, 1996).

14. Rachel Carson, *Silent Spring* (Boston, 1962); Robert Gottlieb, *Forcing the Spring: The Transformation of the American Environmental Movement* (Washington, D.C., 1993),

207-34; Carolyn Merchant, "Earthcare: Woman and the Environmental Movement," *Environment*, 23 (June 1981): 7-10; Grinder, "The Battle for Clean Air," 88-92. On Rachel Carson and the political effect of *Silent Spring*, see Linda Lear, *Rachel Carson: Witness for Nature* (New York, 1997), 404-7, 410-26.

15. On the Big Blackfoot River and Yellowstone National Park threat, see Richard Manning, *One Round River: The Curse of Gold and the Fight for the Big Blackfoot* (New York, 1997), and Tom Hilliard, "Mine, All Mine: The Great Mineral Grab," *The Progressive*, 58 (November 1, 1994). *United States v. Atlantic Richfield Company et al.*, No. CV-89-039-BU-PGH.

16. On coal production and its relationship to air quality legislation, see David McDermott, "Coal Mining in the U.S. West: Price and Employment Trends," *Monthly Labor Review*, 120 (August 1997): 18-24. For a recent critique of the use of "trading rights" in pollution control, see Samuel P. Hays, "Demythologizing the Mythology of Emissions Trading," *Environmental Forum* (January/February 1995): 15-20.

17. Michael P. Malone and Richard B. Roeder, *Montana: A History of Two Centuries* (Seattle, 1976); Michael P. Malone, *Battle for Butte: Mining and Politics on the Northern Frontier, 1864–1906* (Seattle, 1981); Jerry Calvert, *The Gibraltar: Socialism and Labor in Butte, Montana, 1895–1920* (Helena, Mont., 1988); David Emmons, *The Butte Irish: Class and Ethnicity in an American Mining Town, 1875–1925* (Urbana, Ill., 1989); Mary Murphy, *Mining Cultures: Gender, Work, and Leisure in Butte, 1914–41* (Urbana, Ill., 1997).

Chapter 1: A Struggle for Health, Property, and Conservation

1. For more on what economists have to say about pollution, see pages 7.17 to 7.25 of the *Montana Economic Study 1970* directed by Professor Chase of the Bureau of Business and Economic Research at the University of Montana for the Montana Department of Planning and Economic Development and the State Water Resource Board.

2. Charles H. Fulton, "Metallurgical Smoke," *Bureau of Mines Bulletin* 84 (1915), 17.

3. In 1915 the Amalgamated would be dissolved under intense federal pressure and the name Anaconda adopted for the entire organization. See Merrill G. Burlingame and K. Ross Toole, *A History of Montana*, 3 vols. (New York, 1957), 1:279 n. 6, and Toole, *Twentieth-Century Montana*, 122.

4. [Theodore Roosevelt], *The Works of Theodore Roosevelt*, 20 vols. (New York, 1926), 20:393.

5. Granville Stuart, *Pioneering in Montana*, ed. Paul C. Phillips (Lincoln, Nebr., 1977), 229.

Chapter 2: "The Humblest Citizen of Butte is Entitled at Least to Fresh Air"

1. See Ralph I. Smith, "History of the Early Reduction Plants of Butte, Montana," *De Re Metallica*, 18 (May 1953): 1-19; Burlingame and Toole, *History of Montana*,

1:189-90; M. A. Leeson, *A History of Montana, 1739–1885* (Chicago, 1885), 925.

2. *Anaconda Standard*, December 31, 1891 (hereafter *Standard*); Leeson, *A History of Montana*, 917; Smith, "Reduction Plants," 11; *Butte Daily Miner*, July 14, 1885 (hereafter *Miner*).

3. Franklin Farrell to S. T. Hauser, March 1877, Samuel T. Hauser Papers, Montana Historical Society, Helena.

4. This is the core of the language and message of Clark's and Warren's centennial speeches of 1876. See William A. Clark, "Centennial Address," and Charles S. Warren, "Historical Address: The Territory of Montana," in *Contributions to the Historical Society of Montana*, 10 vols. (Helena, Mont., 1896–1940), 2:45-76.

5. R. G. Raymer, *Montana: The Land and the People*, 3 vols. (Chicago, 1930), 1:442-43. Raymer reproduced the figures of the *U.S. Assay Report on Montana, 1882–1889*.

6. A good example of this trend is the Butte and Boston Mining Company, which was formed in 1888 by Andrew Jackson Davis, a Butte mine operator and banker, and a group of Boston capitalists. See Donald MacMillan, "Andrew Jackson Davis: A Story of Frontier Capitalism, 1864–1890" (master's thesis, Montana State University, Missoula, 1958), 41-45.

7. *Engineering and Mining Journal*, January 2, 1892.

8. James Fergus to Andrew Fergus, August 21, 1889, James Fergus Papers (hereafter Fergus Papers), K. Ross Toole Archives, Mansfield Library, University of Montana, Missoula. The six smelters were, running east to west, the Boston and Montana Upper Works, the Boston and Montana Lower Works, the Butte and Boston Company smelter, the Parrott Silver and Copper Company smelter, William A. Clark's Butte Reduction Works, and the Colorado Smelting and Mining Company smelter. Smith, "Reduction Plants," 8-10. Smith provides pictures and a map showing the smelters' locations.

9. Kenneth W. Nelson, "Nonferrous Metallurgical Operations," in *Air Pollution*, ed. Arthur C. Stern, 3 vols. (New York, 1968), 3:171, 173.

10. *Engineering and Mining Journal*, January 24, 1891. See also Smith, "Reduction Plants," 8-11; Burlingame and Toole, *History of Montana*, 1:444, 448; *Standard*, November 29, 1891.

11. *Standard*, November 16, 1890; November 21, 1891.

12. On one occasion the superintendent of the water company reported his "day men had got lost in the smoke and did not reach their work until the forenoon was half gone." *Standard*, January 22, 1891. For other accounts on conditions and incidents, see *Standard*, January 12, 19, 25, February 1, December 12, 1891.

13. *Standard*, January 15, 1890.

14. *Proceedings and Debates of the Constitutional Convention, 1889* (Helena, Mont. 1921), 413.

15. Ibid., 754.

16. *Standard*, December 15, 1890; February 1, 1891.

17. Ibid., December 29, 1890.

18. Ibid., November 29, 1890; January 1, 1891. The paper printed the city health report that listed the deaths by occupation. Out of eighty-six deaths ascribed to pneumonia and consumption, twenty-eight were known as miners, eleven were of unknown occupation but most likely were miners, and thirteen were described as "laborers." See also *Silicosis and Metal Mining Conditions: Hearings before the Subcommittee of the Committee on Education and Labor*, U.S. Senate, S. Con. Res. (1936), 6.

19. *Standard*, November 1, 1890.

20. Ibid., December 29, 1890.

21. Howard Mumford Jones, *The Age of Energy, 1865–1915* (New York, 1971), 11. See also Arthur M. Schlesinger, Jr., and Morton White, eds., *Paths of American Thought* (Boston, 1970), chaps. 7, 8. For development of the professions and a different emphasis, see Robert H. Wiebe, *The Search for Order, 1877–1920* (New York, 1967), chap. 5.

22. *Standard*, May 18, 1891. A comparison of the titles of Reverend Rounder's sermons with those of his counterparts is illustrative. Typical of Rounder's titles were "Go to the Chinaman: Consider His Ways and Live," "Now Is the Time to Tackle the Smoke," and "Honest and Progressive Men." Typical of the sermons of other members of the cloth were "Faith in Jesus Christ," "The Prodigal Daughter," and "Faith and Skepticism." See *Standard*, March 15, May 23, May 18, 1891, for Rounder's sermons and February 4, March 10, December 15, 1891, for the others. On the thought and influence of the social gospel, see Wiebe, *Search for Order*, chap. 3; and Blake McKelvey, *The Urbanization of America, 1860–1915* (New Brunswick, N.J., 1963), chap. 11, on its emergence in cities.

23. *Standard*, March 15, 1891. The city's newspapers continually railed against the Chinese as the source of most of the community's problems.

24. Ibid., March 23, 1891.

25. *Standard*, December 29, 1890. At the time, "bigness" was synonymous with a sense of well-being and goodness. For analysis of this ethic, see Jones, *Age of Energy*; Schlesinger and White, *Paths of American Thought*; and Wiebe, *Search for Order*.

26. *Standard*, December 29, 1890. Dr. Robarts considered diphtheria, typhoid, and scarlet fever preventable.

27. Numerous studies have established the relationship between air pollution and respiratory diseases. Among the most prominent are those summarized in Bertram W. Carnow, M. D., et al., "The Role of Air Pollution in Chronic Obstructive Pulmonary Disease," *Journal of the American Medical Association*, 214, no. 5 (1970), 894, which discusses how air pollution causes pulmonary diseases, heart disease, stroke, carcinoma, chronic bronchitis, and emphysema.

In Bertram W. Carnow, M.D., "Air Pollution and Physician Responsibility," *Archives of Internal Medicine*, 18 (1969), 768-76, Carnow discusses the decrease in infectious diseases and the increase of noninfectious diseases. He summarizes health studies from all over the world that show that air pollution does have an adverse effect on the health of humans, especially those with chronic diseases. See also E. J. Cassell et al., "Air Pollution, Weather, and Illness in a New York Population," *Archives of Internal Medicine*, 18 (1969),

523-30; I. T. T. Higgins, "Effects of Sulfur Oxides and Particulates on Health," *Archives of Environmental Health*, 22 (1971), 584-90; and B. Paccagnella et al., "The Immediate Effects of Air Pollution on the Health of School Children in Some Districts of Ferrara, Italy," *Archives of Environmental Health*, 18 (1969), 495-502.

28. *Standard*, December 6, 29, 1890.

29. Ibid., April 2, 1891.

30. Ibid., December 10, 29, 1890. Robarts considered wage earners "poor people."

31. The editor of the *Standard* was John H. Durston, who had established the paper in 1889 at the request of Marcus Daly of Anaconda. Daly wanted a newspaper to use as an instrument in his developing "war" with William A. Clark for control of Montana politics. Daly threw open his bank account to Durston and told him to build the "best newspaper that could be made."

Publisher Daly allowed editor Durston a free hand editorially, so long as he promoted Daly's political concerns. The two men came to be very close, and during Daly's tenure at Anaconda Copper, Durston was a major figure in the company's chain of command.

The *Standard's* editorial position in the campaign to abate smoke in Butte and the sentiments expressed on the various issues related to it were honestly held. Integrity notwithstanding, there were several reasons the *Standard* assumed an aggressive and almost unequivocating position. As a result of Daly's liberal financing, the paper had a degree of independence not available to the *Butte Inter Mountain* (hereafter *Inter Mountain*), which was a commercial enterprise. In addition, Anaconda did not suffer like Butte did. During much of the period in which the conflict took place, Daly's smelter in Anaconda was either closed down or operating at a reduced rate, sometimes because of a slack copper market, at other times because of remodeling work. For several months in 1891 it was closed as a result of a rate dispute with the Montana Union Railroad, which hauled the ore from Butte to Anaconda. But most importantly, Daly did not engage in heap roasting, his smelters were not situated below the town, and they had high stacks. The prevailing atmospheric conditions in Anaconda were different from those of Butte, and the *Standard* concluded that the stacks were responsible for the comparative lack of smoke in the city. (See a discussion of these conditions in chap. 5.)

Durston followed the antismoke campaign on a day-to-day basis. The *Inter Mountain* could not afford to devote as much space to it as did the *Standard*. The other paper in Butte, the *Miner*, was a weekly owned by W. A. Clark. It ignored the conflict as much as possible, concentrating on local and state politics and on developments in the local mining industry.

Ironically, while the *Standard* championed Butte's cause in this early phase, in the conflicts that followed it would assume a different role. The change in editorial policy came with the change in financial control. In 1899 Daly's Anaconda Copper Company was bought by the Amalgamated—Standard Oil et al.—and Daly died in 1900. Within the year, Durston's influence waned, and he finally was eased out. (The *Standard's* role in succeeding conflicts between the industry and the farmers and the federal government is developed in later chapters.)

For John H. Durston, see *Who Was Who in America*, 8 vols. (Chicago, 1943), 1:350; John Palmer Fought, "Launching the *Standard*," chap. 3 in "John Hurst Durston, Editor, *The Anaconda Standard* in the Clark-Daly Feud" (master's thesis, University of Montana, Missoula, 1959); and the file on Durston and the *Standard* at the School of Journalism, University of Montana, Missoula.

32. *Inter Mountain*, November 20, 1890.

33. *Standard*, November 12, 1890.

34. *Engineering and Mining Journal*, December 20, 1890; *Standard*, November 27, 1890.

35. *Standard*, March 12, 1891; *Engineering and Mining Journal*, January 10, 1891; *Standard*, January 3, 1891. From time to time, the *Standard* published solutions received at its office. Most accounts assured success at very little cost but contained crude explanations and little detail. For conversion to sulfuric acid, many suggested a process similar to the modern method called the lead chamber process. A "scrubbing" device is generally any gas-cleaning device where the separated gas, vapor, or particulate matter leaves the process dissolved or suspended in a liquid. The scrubbing liquid is usually water, but it may be an aqueous solution of neutralizing chemicals or any liquid that proves effective for the specific application. See Stern, *Air Pollution*, 2:333.

36. *Standard*, November 21, 26, 27, 1890.

37. Ibid., December 6, 1890.

38. For the various solutions and discussions, see the *Standard*, November 21, 26, 27, 1890.

39. Ibid., November 27, 1890.

40. Ibid.

41. Ibid., November 21, 1890.

42. *Inter Mountain*, November 20, 1890; *Standard*, November 21, December 3, 5, 7, 8, 1890; March 2, 1891; Smith, "Reduction Plants," 12. Actually the Boston and Montana was already planning to move to Great Falls, Montana, because of the increasing shortages of water, power, and room for expansion in the Butte area and because of the proximity of the coalfields of Sand Coulee to Great Falls. The Butte and Boston had similar plans for the same reasons. In 1895 the two corporations would be consolidated.

43. *Standard*, December 4, 18, 1890. The ordinance was modeled after the Chicago smoke ordinance of 1881, the first air pollution law in the United States. The smoke committee had urged copying the law. On the Chicago ordinance, see Jack Bregman and Sergei Lenormond, *The Pollution Paradox* (New York, 1966), 6.

44. The effective dispersion of stack effluents is an engineering art in itself, requiring sophisticated knowledge of geometry, chemistry, and meteorology. See Stern, *Air Pollution*, 1:chap. 6.

45. The *Industrial World* article was reprinted in the *Standard*, December 29, 1890.
46. *Standard*, November 27, December 6, 11, 1890.
47. Ibid., November 27, December 6, 7, 10, 1890.
48. Ibid., December 16, 21, 1890; February 2, 1891.
49. Ibid., December 16, 1890.
50. Ibid., December 17, 25, 1890.
51. Ibid., December 16, 1890.
52. Ibid., December 17, 1890.
53. Ibid., December 29, 1890.

Chapter 3: "Bluffs Don't Go in Smoke Wars"

1. Observations were made at 9 A.M., 2 P.M., and 8 P.M. daily and recorded by the health department. The *Standard* published the department's report each week. See the issues of January 12, 18, 19, February 1, 1891.
2. Ibid., January 22, 1891.
3. Ibid.; *Engineering and Mining Journal*, 51:150; *Standard*, January 24, April 2, 1891. Dr. Robarts's assessment of Hutchinson's claims is not as rash as it may appear. Chemists of the time lacked knowledge of the size of atoms, of their structures, and of some of the other properties associated with them. See Lawrence Eblin, *Chemistry: A Survey of Fundamentals* (Palo Alto, Calif., 1968), chap. 3.
4. *Standard*, January 25, 1891.
5. Ibid., January 24, 25, 1891.
6. Ibid. January 25, 1891.
7. Ibid.
8. Ibid.
9. Ibid.
10. Ibid., January 29, 1891; *Miner*, January 29, 1891.
11. *Standard*, January 29, 1891.
12. Ibid.
13. Ibid.
14. *Engineering and Mining Journal*, February 1891.
15. Ibid.
16. Ibid.; *Standard*, January 29, 1891.
17. *Standard*, January 29, 30, 1891.
18. Ibid., February 2, 1891.
19. Ibid.
20. Ibid.
21. Ibid., February 27, May 18, 1891.
22. Ibid. See also the comments in issues of January 21, 30, February 5, 9, 1891.
23. *Standard*, February 11, 17, 27, March 6, 1891.

24. Ibid., February 21, 1891.
25. Ibid., February 23, 26, 1891.
26. Ibid., February 23, 27, 1891.
27. Ibid., March 4, 5, 6, 1891.
28. Ibid., March 7, 8, 9, 1891.
29. See *Standard*, March 4, 6, 1891.
30. Ibid., January 1, 1892. This issue provided a breakdown of recorded deaths on a monthly basis for 1891. Total number of deaths was listed as 620. Ibid., February 13, March 6, April 4, 1891.
31. Ibid., March 9, 11, 1891; November 21, 1890.
32. Ibid., March 11, 14, 1891.
33. Ibid., March 9, 7, 1891.
34. Robert G. McCloskey, *The American Supreme Court* (Chicago, 1960), 167.
35. *Miner*, March 12, 1891; *Standard*, March 7, 1891.
36. Ibid., March 9, 1891.
37. Ibid.
38. Ibid.
39. Ibid., May 18, March 21, 22, 1891.
40. The chemical formula for the process as reported in the *Standard* was $1SO_2 + 2H_2O \rightarrow NaSO_4 + H + Cl$. The formula was reportedly chemically feasible, but costs were dependent on how much sulfate of soda ($NaSO_4$) was lost in the process of conversion. Analysis by Phil Torangeau, Chemistry Department, University of Montana, Missoula.
41. *Standard*, March 24, 25, 26, 1891.
42. Ibid., April 2, 1891.
43. Ibid.

Chapter 4: "The War of Wealth against Health"

1. *Standard*, April 1, 15, 18, 23, 1891. Storer was forced to return to New York.
2. Ibid., March 23, 1891.
3. Ibid., May 18, 1891.
4. Ibid.
5. *Inter Mountain*, November 20, 26, 1890.
6. Ibid., November 26, 28, 1890.
7. See the accounts in the local newspapers in late April 1891 that depict the agonizing of city officials over establishing and funding the city budget. The alleged ability of Democrats as opposed to Republicans to attract foreign capital was a prominent plank in the Democratic platform for the city elections of 1891. See the *Standard*, April 8, 1891. On how hungrily the city sought eastern capital, see K. Ross Toole's "When Big Money Came to Butte: The Migration of Eastern Capital to Montana," *Pacific Northwest Quarterly*, 44 (January 1953), 27-49.

8. *Engineering and Mining Journal,* January 2, 1892, 58.

9. Quoted from Eric Goldman, *Rendezvous with Destiny: A History of Modern American Reform* (New York, 1952), 3.

10. *Weekly Missoulian,* November 11, 1891.

11. *Standard,* December 5, 8, 10, 17, 1891.

12. Raymer, *Montana,* 1:451; *Standard,* December 10, 1891; February 12, 1892.

13. *Standard,* November 29, December 6, 1891.

14. Ibid., December 9, 1891.

15. Ibid., December 5, 1891.

16. Ibid.

17. Ibid., December 8, 9, 1891.

18. Ibid., December 9, 1891.

19. Ibid., December 8, 9, 1891.

20. Ibid., December 10, 1891.

21. Ibid., December 14, 1891; *Inter Mountain,* December 15, 1891. See also the comments of prominent residents made during a citizens' meeting and printed in the *Standard,* December 14, 1891.

22. *Standard,* December 11, 1891.

23. Ibid., December 12, 1891.

24. Ibid.

25. Ibid., December 14, 1891.

26. *Standard,* December 14, 1891. For biographical information on Irvine, see Joaquin Miller, *An Illustrated History of Montana* (Chicago, 1894), 771.

27. *Standard,* December 14, 1891.

28. Ibid., December 15, 1891.

29. This idea may have originated with the *Inter Mountain.* When appeals to the local officials of the Boston and Montana proved fruitless, the *Inter Mountain* featured an editorial recommending the city appeal to the wisdom of the company's top executives in Boston. See the *Inter Mountain,* December 15, 1891. The telegrams were reprinted in the *Standard,* December 15, 1891.

30. *Standard,* December 15, 1891.

31. Ibid. The men underwriting the bond were Charles S. Warren, John Forbis, James Forbis, H. G. Valiton, William Owsley, A. J. Davis, J. A. Talbot, A. W. Barnard, H. L. Frank, and William L. Hoge. See the *Standard,* December 15, 1891; Miller, *Illustrated History,* passim; and *Progressive Men of Montana* (Chicago, 1902), passim, for their backgrounds.

32. *Standard,* December 15, 1891; Miller, *Illustrated History,* 253.

33. *Standard,* December 15, 1891.

34. Ibid., December 16, 17, 1891.

35. Ibid., December 13, 17, 1891. The paper reprinted the court transcript.

36. Ibid., December 17, 1891.

37. Ibid.

38. Ibid.

39. Ibid., December 16, 18, 1891.

40. Ibid., December 17, 19, 1891.

41. Ibid., December 22, 1891.

42. Ibid.

43. Ibid.

44. Ibid., December 20, 22, 27, 28, 1891.

45. Ibid., December 22, 1891.

46. Ibid.

47. See the explanation presented by the representative of the Anaconda Copper and Silver Mining Company at the citizens' meeting of November 20, 1890, and printed in the *Standard*, November 21, 1890.

48. Raymer, *Montana*, 1:450. From 104.6 million tons at the end of 1889 to 164 million in 1892. For the other statistics, see Raymer, *Montana*, 1:484.

49. James Fergus to Andrew Fergus, February 13, 1901, Fergus Papers.

50. Smith, "Reduction Plants," 14-17.

Chapter 5: Progress and Pollution Come to the Valley

1. Burlingame and Toole, *History of Montana*, 2:199.

2. Raymer, *Montana*, 1:449; "Notes on the Metallurgy of Copper of Montana," *Transactions of the American Institute of Mining Engineers* (hereafter *AIME*), 34 (1904), 265-67; R. E. Swain and W. D. Harkins, "Arsenic in Vegetation Exposed to Smelter Smoke," in *Papers on Smelter Smoke and Arsenical Poisoning* (n.p., 1908), 915-28, copy in Mansfield Library, University of Montana, Missoula (hereafter *Papers on Smelter Smoke*).

3. Among the more valuable entities included were the Parrot Silver and Copper Company, the Washoe Copper Company, the Colorado Smelting and Mining Company, the Diamond Coal and Coke Company, and the Boston and Montana Company. See *Engineering and Mining Journal*, 67:630; Thomas W. Lawson, *Frenzied Finance* (New York, 1905), 237-52; K. Ross Toole, "The History of the Anaconda Copper Mining Company: A Study in the Relationship between a State, Its People, and a Corporation" (Ph.D. diss., University of California, Los Angeles, 1954), 105-7, 116-18; Burlingame and Toole, *History of Montana*, 1:206-7.

4. *186 Federal Reporter*, (St. Paul, Minn., 1951), 795; "Notes on the Metallurgy of Copper," 267.

5. Swain and Harkins, "Arsenic in Vegetation," 916; W. C. Staton to Theodore Roosevelt, December 3, 1908, Department of Justice File 144276 (hereafter DJ File 144276), National Archives, Washington D.C.

6. Swain and Harkins, "Arsenic in Vegetation," 915.

7. Ibid., 916; W. D. Harkins and R. E. Swain, "The Chronic Arsenical Poisoning of

Herbivorous Animals," in *Papers on Smelter Smoke,* 928-29; Brief and Argument for Appellant, *Fred J. Bliss v. Washoe Copper Company et al.*, vol. 8, United States of America in the U.S. Circuit Court of Appeals, for the Ninth Circuit (hereafter Brief and Argument for Appellant), DJ File 144276.

8. Harkins and Swain, "Chronic Arsenical Poisoning," 929.

9. Affidavit of Peter Levengood (hereafter Levengood affidavit), DJ File 144276.

10. Affidavit of N. J. Bielenberg (hereafter Bielenberg affidavit), DJ File 144276.

11. Swain and Harkins, "Arsenic in Vegetation," 917; Harkins and Swain, "Chronic Arsenical Poisoning," 932-34, 939; C. Ackley, F. R. Holden, and P. L. Magill, eds., *Air Pollution Handbook* (New York, 1956), 83, 84.

12. Brief and Argument for Appellant.

13. Ibid., 119-21; W. C. Staton to Theodore Roosevelt, January 18, 1909, DJ File 144276.

14. Bielenberg affidavit.

15. *Standard,* March 9, 1891.

16. *Engineering and Mining Journal,* 72:505, 73:191; "Notes on Metallurgy of Copper," 267.

17. *Standard,* May 5, 1908; *Engineering and Mining Journal,* 72:505, 73:191; *186 Federal Reporter,* 803; Raymer, *Montana,* 1:501.

18. Ligon Johnson to Attorney General, Washington, D.C., DJ File 144276. Italics in original.

19. Edgar M. Dunn, "Determination of Gases in Smelter Flues; and Notes on the Determination of Dust Losses at the Washoe Reduction Works, Anaconda, Montana," *AIME,* 46 (1913), 666; James O. Elton, "Arsenic Trioxide from Flue Dust," *AIME,* 46 (1913), 691.

20. *186 Federal Reporter,* 803.

21. Fulton, "Metallurgical Smoke," 38.

22. Ibid., 30.

23. Quoted from Staton to Roosevelt, January 18, 1909; see also Brief and Argument for Appellant, 122; Bielenberg affidavit; Harkins and Swain, "Chronic Arsenical Poisoning," 929.

24. *186 Federal Reporter,* 803; Swain and Harkins, "Arsenic in Vegetation," 915-16.

25. Harkins and Swain, "Chronic Arsenical Poisoning," 930-31, 938-39.

26. Ibid., 930-31, 945; Brief and Argument for Appellant.

27. *Standard,* November 10, 1904.

28. Brief and Argument for Appellant.

29. W. D. Harkins and R. E. Swain, "The Determination of Arsenic and Other Solid Constituents of Smelter Smoke, with a Study of the Effects of High Stacks and Large Condensing Flues," in *Papers on Smelter Smoke,* 995-96. Italics added.

30. Ibid., 996.

31. Ibid., 973.

32. Fulton, "Metallurgical Smoke," 16, 24, 38-39, 81. Italics added.

33. For the various statistics, see Harkins and Swain, "Determination of Arsenic and Other Solid Constituents," 992-95.

34. Swain and Harkins, "Arsenic in Vegetation," 917-18.

35. Brief and Argument for Appellant.

36. *Standard*, September 12, 1906; Brief and Argument for Appellant.

Chapter 6: The Farmers versus the Trust

1. From a letter of the Deer Lodge Valley Farmers' Association to the Anaconda Copper Mining Company, March 4, 1905, printed in *186 Federal Reporter*, 817-19; see also Harkins and Swain, "Chronic Arsenical Poisoning," 927; and Fulton, *Metallurgical Smoke*, 31.

2. Staton (chairman of the executive committee of the Deer Lodge Valley Farmers' Association) to Roosevelt, December 3, 1908; *Standard*, October 3, 1909.

3. *186 Federal Reporter*, 789-92.

4. Ibid.

5. Ibid., 790-91.

6. Brief and Argument for Appellant, 2351-52; *Inter Mountain*, May 19, 1909.

7. The defendant's answer to the Bliss complaint was printed in the Anaconda *Standard*, n.d. The clipping is contained in a scrapbook of newspaper clippings about Montana politics and the Anaconda "smoke case" in the Montana Room of the Mansfield Library, at the University of Montana, Missoula. See also *186 Federal Reporter*, 794-95, 803.

8. *186 Federal Reporter*, 794-95, 803.

9. Ibid.; *Standard*, September 12, 1906.

10. *Inter Mountain*, May 19, 1909.

11. *Standard*, November 10, 1906; Fulton, "Metallurgical Smoke," 83; Harkins and Swain, "Chronic Arsenical Poisoning," 946.

12. *Standard*, November 10, 1906.

13. Brief and Argument for Appellant, 2368.

14. See Richard Hofstadter, *Anti-intellectualism in American Life* (New York, 1962), 208-11, 209n.

15. *Inter Mountain*, May 19, 1909; Brief and Argument for Appellant, 137.

16. Brief and Argument for Appellant, 124.

17. Swain and Harkins, "Arsenic in Vegetation," 923, 920.

18. Ibid., 921-23, 926.

19. Harkins and Swain, "Chronic Arsenical Poisoning," 937-42.

20. Brief and Argument for Appellant, 136-37.

21. *Standard*, September 11, 1906; Brief and Argument for Appellant, 2353.

22. *Standard*, September 11, 1906, December 5, 1908.

23. Ligon Johnson to Charles J. Bonaparte, January 25, 1909; Charles Doremus to

C. F. Kelley, January 25, 1909; Brief and Argument for Appellant, 134-35; all in DJ File 144276.

24. Johnson to Bonaparte, January 25, 1909; Charles J. Bonaparte to Theodore Roosevelt, January 27, 1909; Ligon Johnson to W. J. Hughes, July 20, 1910; all in DJ File 144276; *Standard*, February 15, 1909; *Inter Mountain*, February 17, 1909.

25. Johnson to Bonaparte, January 25, 1909; Testimony of Prof. F. W. Traphagen, DJ File 144276.

26. Brief and Argument for Appellant, 2356-57.

27. *Standard*, December 27, 1907; Brief and Argument for Appellant, 2361; *186 Federal Reporter*, 826.

28. Staton to Roosevelt, January 18, 1906.

29. Brief and Argument for Appellant, 132-33.

30. Ibid., 2358-59.

31. Ibid., 2361.

32. Ibid., 2360.

33. Brief and Argument for Appellant, 2360-61; Johnson to Bonaparte, January 25, 1909. Copy of Dr. Gardiner's testimony is in DJ File 144276.

34. George W. Wickersham to Dr. J. A. Holmes, August 28, 1911, DJ File 144276.

35. Testimony of Prof. Joseph Blankenship, DJ File 144276.

36. Harkins and Swain, "Chronic Arsenical Poisoning," 942; Brief and Argument for Appellant, 2370-71.

37. Brief and Argument for Appellant, 130-31.

38. Ibid., 132-33, 2351, 2372, 2379.

39. Ibid., 124-25.

40. *186 Federal Reporter*, 790, 814.

41. Ibid., 826-27; 820, 823-24, 804, 790.

42. Staton to Roosevelt, January 18, 1909.

43. *Inter Mountain*, May 26, 1909.

Chapter 7: The Struggle outside the Court

1. *Standard*, March 7, 1911.

2. Howard, *Montana*, 69.

3. See Toole, *Montana*, 186-211; Lawson, *Frenzied Finance*, 4 ff.; C. B. Glasscock, *The War of the Copper Kings* (New York, 1935), 259, passim; Sarah McNelis, *Copper King at War: The Biography of F. Augustus Heinze* (Missoula, Mont., 1968); C. P. Connolly, *The Devil Learns to Vote* (New York, 1938); K. Ross Toole, "Marcus Daly: A Study of Business in Politics" (master's thesis, Montana State University, 1948); and "The Genesis of the Clark-Daly Feud," *Montana Magazine of History*, 2 (April 1951), 21-33; and Howard, *Montana*.

4. See testimony of Dr. O. Y. Warren and W. C. Staton to Ligon Johnson, January 5,

1909, DJ File 144276; Brief and Argument for Appellant, 2341.

5. Testimony of Dr. Salmon, DJ File 144276; and testimony quoted in Brief and Argument for Appellant, 2353, 2346.

6. Brief and Argument for Appellant, 137, 2350-51; Staton to Roosevelt, December 3, 1908.

7. For interpretive studies on the Montana press, see Toole, *Twentieth-Century Montana*, chaps. 4, 7, 11; John M. Schiltz, "Montana's Captive Press," *Montana Opinion*, 1 (June 1956), 1-12; Richard T. Ruetten, "Anaconda Journalism: The End of an Era," *Journalism Quarterly* (Winter 1960), 3-13.

8. *Standard*, June 11, 1906.

9. Ibid., September 12, 1906.

10. Ibid., October 6, 1907.

11. Ibid., March 7, 1911.

12. Ibid., August 20, 1905.

13. Staton to Roosevelt, December 3, 1908.

14. Harry J. Quinlan to Woodrow Wilson, July 21, 1913, DJ File 144276.

15. (Deer Lodge) *Powell County Call*, January 27, 1906.

16. *Standard*, April 30, 1908.

17. *Inter Mountain*, October 4, 1905.

18. *Standard*, October 3, 1907.

19. Ibid., January 5, 7, 1909.

20. Staton to Roosevelt, December 3, 1908.

21. Staton to Roosevelt, January 18, 1909.

22. *Standard*, January 7, 9, February 16, 1909; Staton to Roosevelt, January 18, 1909.

23. Brief and Argument for Appellant, 2362-66.

24. Ibid., 2363.

25. George F. Peirce, R. E. Swain, and J. Pearce Mitchell, Investigation and Report to Ligon Johnson, October 21, 1910, DJ File 144276. Peirce, Swain, and Mitchell were chemists and botanists from Stanford University.

26. There is no definitive work on Thomas Carter. The best profile of the man can be obtained by reading his papers, the Thomas H. Carter Papers (hereafter Carter Papers), on file at the Library of Congress, Washington, D.C. A fair picture of the man's political life can be pieced together by reading Elmo R. Richardon's *The Politics of Conservation, 1897–1913* (Berkeley, Calif., 1962), and Gifford Pinchot's *Breaking New Ground* (New York, 1947).

27. *Congressional Record*, 60th Cong., 1st sess., 1908, 42, pt. 7:6238; Staton to Roosevelt, January 18, 1909.

28. Staton to Roosevelt, January 18, 1909. One definitive work on the career of Joseph M. Dixon is Jules A. Karlin, *Joseph M. Dixon of Montana* (Missoula, Mont., 1974). A good picture of the man's career can also be obtained by consulting Jules

Karlin, "Progressive Politics in Montana," in Burlingame and Toole, *History of Montana*, 1:chap. 11; Toole, *Montana*, chap. 10; and Toole, *Twentieth-Century Montana*, chaps. 9, 10, 296-97.

29. See unidentified newspaper clipping dated November 14, 1908, attached to letter of Theodore Roosevelt to Charles J. Bonaparte, December 9, 1908, and Ernest Beaudette to William Taft, May 5, 1911; both in DJ File 144276.

30. See Theodore Roosevelt to Conrad Kohrs, September 9, 1908, in *The Letters of Theodore Roosevelt,* ed. Elting E. Morison, 6 vols. (Cambridge, Mass., 1952), 6:1212-18.

31. Quoted from John M. Blum, "The Republican Roosevelt," in *Theodore Roosevelt: A Profile*, ed. Morton Keller (New York, 1967), 177.

32. Ibid., 178. The interpretation of Roosevelt's domestic, political, and economic policy relies on Blum's analysis of Roosevelt in the above mentioned article and in his *The Republican Roosevelt* (Cambridge, Mass., 1954). Roosevelt's letters and his role in the Deer Lodge farmers' struggle with Amalgamated and the subsequent conflict between his administration and Amalgamated confirms Blum's view of the nature of Roosevelt's leadership and national objectives.

33. Staton to Roosevelt, January 18, 1909.

34. Ibid., Staton to Roosevelt, December 3, 1908.

35. Ibid.

36. Ibid., January 18, 1909.

37. Ibid., December 3, 1908.

Chapter 8: The Roosevelt Men versus the Smelters

1. The conflict between the federal government and the trust emerged and, for the most part, worked itself out within the social, political, and economic milieu of the Progressive Era. The progressive movement, although difficult to describe or define precisely, covered a period in which the urge for reform dominated American thinking and politics. Fundamentally, progressivism was a reaction to the course America had taken in the two generations that followed the Civil War. It was a reaction to the emergence of giant economic entities that dominated not only American economic life but its political life as well. Reformers of the period sought to prevent the collective forces spawned by big business, organized labor, and implacable political machines from crushing the individual. Richard Hofstadter terms the movement the "complaint of the unorganized against the consequences of organization." Hofstadter, *Age of Reform: From Bryan to F.D.R.* (New York, 1955), 214.

2. For the history and analyses of these conflicting and competing philosophies, see Sidney Fine, *Laissez Faire and the General-Welfare State: A Study of Conflict in American Thought, 1865–1901* (Ann Arbor, Mich., 1956); Richard Hofstadter, *Social Darwinism in American Thought* (Boston, 1955); Robert Green McCloskey, *American Conservatism in the Age of Enterprise, 1865–1910* (New York, 1951); Hofstadter, *The Age of Reform*;

Goldman, *Rendezvous with Destiny*; Harry K. Girvetz, *The Evolution of Liberalism* (New York, 1963); "Revolution in Values," chap. 6 in Wiebe, *Search For Order*; Max Lerner, "The Triumph of Laissez-Faire," in Schlesinger and White, *Paths of American Thought*, 147-66; and Paul A. Samuelson, "Economic Thought and the New Industrialism," in Schlesinger and White, *Paths of American Thought*, 204-18.

3. Morrison, *Letters of Theodore Roosevelt*, 6:1374; *Works of Theodore Roosevelt*, 15:493; Wiebe, *Search for Order*, 189-93; Blum, "Republican Roosevelt," 187-89; George E. Mowry, *Era of Theodore Roosevelt and the Birth of Modern America, 1900–1912* (New York, 1958), chap. 11.

4. Wiebe, *Search for Order*, 190; Blum, "Republican Roosevelt," 186. In a letter to William Howard Taft, President Roosevelt discussed some of the major problems facing the country and the role of the presidency in contending with them. As to the presidency, he said: "Now I think all these tendencies for evil can be overcome, No one man and no two men can do very much working alone to overcome them. But simply because of his position, the man who is President can do most if he chooses to try to do it." See Theodore Roosevelt Papers (hereafter TR Papers), microfilm series 2, reel 353, vol. 89, letter 135, Library of Congress, Washington, D.C. See also Wiebe's analysis of the rising prominence of the executive and Roosevelt's effect on it (Wiebe, *Search for Order*, 188-90). For insights into the personal relationships and shared objectives of Roosevelt's men, see Theodore Roosevelt to Gifford Pinchot, TR Papers, microfilm series 2, reel 354, vol. 91, letter 366, in which Roosevelt talks about their causes and about the shared intimacy of the two correspondents and James Garfield, secretary of the interior; Charles J. Bonaparte to Theodore Roosevelt, December 2, 1908, Charles J. Bonaparte Papers (hereafter Bonaparte Papers), Library of Congress, Washington, D.C., in which Bonaparte discusses the administration's views on "the Interests," and in the same collection, filed under "Subject File," Bonaparte's "Statement of Principles" for the National Conservation Association. See also Pinchot, *Breaking New Ground*. For an analysis of the quality of the administration and its members, see Elmo H. Richardson, *Politics of Conservation, 1897–1913* (Berkeley, Calif., 1962), 16-34 *ff*; and Wiebe, *Search for Order*, 189-92.

5. Pinchot, *Breaking New Ground*, 464, 190; Mowry, *Era of Theodore Roosevelt*, 214.

6. See Roosevelt to James Wilson, June 7, 1907, TR Papers, microfilm series 2, reel 345, vol. 73, for a comprehensive statement of the president's conservation philosophy. The letter comprises seven pages.

7. Richardson, *Politics of Conservation*, 21; *Works of Theodore Roosevelt*, 15:517-23.

8. Ligon Johnson to H. M. Hoyt, March 19, 1908, DJ File 144276.

9. Ligon Johnson to Charles J. Bonaparte, November 7, 1907, DJ File 144276; J. K. Haywood, "Injury to Vegetation and Animal Life by Smelter Wastes," U.S. Department of Agriculture, *Bureau of Chemistry Bulletin No. 113* (1910); Fulton, "Metallurgical Smoke." See also the letter of Ligon Johnson to W. J. Hughes, July 3, 1913, DJ File 144276.

10. J. K. Haywood, "Injury to Vegetation by Smelter Fumes," U.S. Department of

Agriculture, *Bureau of Chemistry Bulletin No. 89* (1905), 1-21.

11. Haywood, "Injury . . . by Smelter Wastes."

12. Ibid., 21-36.

13. Fred Morrell to William B. Greeley, January 14, 1924, DJ File 144276. The district forester herein reviewed the studies made from 1905 to 1910 in the areas of California and Montana that were subject to smelter fumes.

14. Charles J. Bonaparte to Ligon Johnson, November 13, 1907; Ligon Johnson to Charles J. Bonaparte, November 7, 1907; H. M. Hoyt to Ligon Johnson, June 11, 1908; Ligon Johnson to W. J. Hughes, August 21, 1914; all in DJ File 144276.

15. Johnson also created and led one of the largest conservationist delegations ever to attend a congressional hearing on behalf of the formation of the Appalachian National Park Association. See his letter to Bonaparte, November 24, 1908, DJ File 144276.

16. Johnson to Bonaparte, January 25, 1909.

17. Ibid.

18. For clear statements of this "make sure of victory" attitude, see Roosevelt to Charles J. Bonaparte, August 15, 1908, Bonaparte Papers, "Special Correspondence with Theodore Roosevelt"; Gifford Pinchot to James R. Garfield, May 22, 1909, Gifford Pinchot Papers (hereafter Pinchot Papers), Container 115, 1909, Library of Congress, Washington, D.C.

19. Johnson to Bonaparte, November 7, 1907.

20. Ibid.

21. Ibid.

22. Ibid.

23. Ibid.

24. Ibid. A leading Montana historian pointed out that the "shutdown" was a device the Amalgamated and Anaconda Copper Mining Company resorted to frequently in the early decades of the twentieth century to contend with labor. In describing the "great shutdown" of 1903, K. Ross Toole wrote, "At this juncture the Company employed a tactic it was to use again and again in its relationship with labor throughout the years, a tactic which ultimately became monotonous in its predictability." Toole, *Twentieth-Century Montana*, 89.

25. Johnson to Bonaparte, November 7, 1907.

26. Bonaparte to Johnson, November 13, 1907.

27. See Attorney General's Office to Ligon Johnson, May 29, 1908; Ligon Johnson to W. J. Hughes, June 16, 1908; W. J. Hughes to Ligon Johnson, June 18, 1908; and Department of Justice Memorandum by W. J. Hughes, August 19, 1908; all in DJ File 144276; Haywood, "Injury . . . by Smelter Wastes"; Brief and Argument for Appellant.

28. Hoyt to Johnson, June 11, 1908; Johnson to Hughes, August 21, 1914.

29. Johnson to Bonaparte, November 24, 1908.

30. Ibid.; Memorandum by Attorney General's Office for the Solicitor General, June 12, 1908, DJ File 144276.

31. Johnson to Bonaparte, November 24, 1908; *Intermountain Republican*, July 13, 1908, a newspaper clipping found among the reports submitted to Bonaparte by Johnson on the Tennessee smelters' success with the marketing of fertilizer. DJ File 144276.

32. See Memorandum by Ligon Johnson to the Honorable Attorney General, in re Smelter Fumes Memorandum of Authorities, n.d., DJ File 144276. The memorandum cites fourteen pages of authorities for the government to proceed as sovereign in abating the smelter fumes.

33. W. G. Weigle to J. H. Hughes, August 14, 1908, DJ File 144276. Italics in original.

34. Ibid.

35. Hoyt to Johnson, June 11, 1908; R. L. Clinton to W. J. Hughes, August 4, 1908, DJ File 144276.

36. Clinton to Hughes, August 4, 1908.

37. Memorandum by Solicitor General, August 7, 1908; Hughes, Department of Justice Memorandum, August 19, 1908; Memorandum by W. J. Hughes for the Attorney General, October 21, 1908; all in DJ File 144276.

Chapter 9: Compelling a Corporation to Do Its Duty

1. Hughes, Memorandum for Attorney General, October 21, 1908.

2. See Theodore Roosevelt to Henry Cabot Lodge, October 7, 1908, TR Papers, microfilm series 2; Theodore Roosevelt to Joseph Dixon, November 4, 1908, ibid.; Morison, *Letters of Theodore Roosevelt*, 6:1124.

3. Wiebe, *Search for Order*, 192; Appointment Diary, December 3, 5, 1908, TR Papers, microfilm series 9; Charles J. Bonaparte to Theodore Roosevelt, December 8, 1908, DJ File 144276; Johnson to Bonaparte, November 24, 1908.

4. Appointment Diary, December 5, 1908; *Inter Mountain*, December 6, 1908.

5. Johnson to Bonaparte, November 24, 1908.

6. Stern, *Air Pollution*, 1:222; Johnson to Bonaparte, November 24, 1908; Bonaparte to Roosevelt, December 8, 1908; Memorandum by Ligon Johnson for Mr. Hughes, n.d., DJ File 144276.

7. *Miner*, March 6, 1909. The *Miner* reprinted an article from the *London Mining Journal* (February 20, 1909) titled "Arsenic Fume." The article quoted testimony from the Bliss case. John D. Ryan to Theodore Roosevelt, December 8, 1908; Johnson to Bonaparte, November 24, 1908; both in DJ File 144276; Theodore Roosevelt to Charles J. Bonaparte, December 9, 1908, TR Papers, microfilm series 2. In this letter, Roosevelt discusses the allegations made by Amalgamated.

8. Roosevelt to Bonaparte, December 9, 1908, TR Papers.

9. Memorandum as to Conference with President about Proposed Anaconda Suit by Attorney General Charles Bonaparte, December 8, 1908, DJ File 144276; Harry W. Laidler, *Concentration of Control in American Industry* (New York, 1931), 320-21, 310-

11, 441-43. Laidler shows how extensive such interconnections were and points out that the practice of appointing dummies to represent the more prominent figures in the financial world on boards of directors makes it impossible sometimes to guess at the interests actually represented.

10. Bonaparte, Memorandum as to Conference with President, December 8, 1908.

11. Hoyt to Johnson, June 11, 1908.

12. *In invitum* is a term applied to proceedings against an adverse party, to which the party does not consent. See Henry Campbell Black, *Black's Law Dictionary* (St. Paul, Minn., 1968), 896. For Roosevelt and Bonaparte's discussion, see Bonaparte, Memorandum as to Conference with President, December 8, 1908.

13. Roosevelt to Bonaparte, December 9, 1908, DJ File 144276; Hughes, Department of Justice Memorandum, August 19, 1908; Roosevelt to Bonaparte, December 9, 1908, TR Papers, microfilm series 2.

14. Johnson to Bonaparte, November 24, 1908; Roosevelt to Bonaparte, December 9, 1908, TR Papers, microfilm series 2.

15. Bonaparte, Memorandum as to Conference with President, December 8, 1908.

16. Ibid.

17. See "Report on Forest Conditions in the Vicinity of Anaconda, Montana, as Compared with Those in Adjacent Forests," by George Grant Hedgcock (1910), pathologist in charge, Forest Disease Survey, TR Papers, microfilm series 2.

18. See David Starr Jordan to Gifford Pinchot, telegram, December 7, 1908; George F. Peirce to Theodore Roosevelt, December 15, 1908; Samuel Knight to George F. Peirce, n.d., 1908; George F. Peirce to Samuel Knight, June 25, 1909; all in TR Papers, microfilm series 2.

19. Johnson to Bonaparte, January 25, 1909.

20. Ibid.

21. See Charles A. Doremus to C. F. Kelley, December 21, 1908; Johnson to Bonaparte, January 25, 1909.

22. Johnson to Bonaparte, January 25, 1909.

23. Ibid.

24. Ibid.

25. Ibid.

26. Ibid.

27. Ibid.

28. C. F. Kelley to H. M. Hoyt, January 23, 1909; Charles J. Bonaparte to C. F. Kelley, January 27, 1909; Bonaparte to Roosevelt, January 27, 1909; all in DJ File 144276.

29. Kelley to Hoyt, January 23, 1909. Ryan was, of course, the president of Amalgamated Copper; B. B. Thayer was the president of the Anaconda Copper Mining Company.

30. Bonaparte to Kelley, January 27, 1909.

31. Ibid.

32. Ibid.

33. Bonaparte to Roosevelt, January 27, 1909.

34. Ibid.

35. Ibid.

36. Theodore Roosevelt to Charles J. Bonaparte, February 4, 1909, TR Papers, microfilm series 2.

37. Johnson to Bonaparte, January 25, 1909.

38. H. M. Hoyt to Charles J. Bonaparte, February 16, 1909, DJ File 144276.

39. Ibid.

40. H. M. Hoyt to Charles J. Bonaparte, February 17, 1909, DJ File 144276.

41. Theodore Roosevelt to Charles J. Bonaparte, February 18, 1909, TR Papers, microfilm series 2.

Chapter 10: The Taft Men versus the Smelters

1. See the *Standard* and the *Inter Mountain* on December 4, 5, 6, 7, 8, 1908. The *Great Falls Tribune* (hereafter *Tribune*) carried news accounts of the story, but the reporting was restrained.

2. *Standard*, December 5, 1908; *Inter Mountain*, December 5, 1908; *Tribune*, December 5, 1908.

3. Toole, *Twentieth-Century Montana*, 114-15.

4. *Inter Mountain* and *Standard*, December 5, 1908.

5. *Standard*, December 5, 6, 7, 8, 1908. See especially December 6. Although the various resolutions and telegrams of protest were prominently displayed in the *Standard* and reported in the *Inter Mountain*, few were actually sent to Roosevelt. His secretary kept a careful tally of all incoming mail, and a check of this ledger for the month of December 1908 revealed that only six letters were received from Montanans in reference to the Washoe smelter case. See TR Papers, microfilm series 11, 1903-9.

6. *Standard*, December 26, 1908.

7. Ibid., December 25, 1908; Johnson to Bonaparte, January 25, 1909.

8. *Tribune*, December 9, 1908. Between December 5 and 8 the *Tribune* ran only one relatively short report each day on the story.

9. *Standard*, December 10, 26, 1908.

10. Ibid., December 25, 1908. See also items in the Butte papers on the same theme. The *Tribune*, December 7, 1908, published a long editorial on the front page from the editor of the *New York Sun* that was sarcastically critical of Roosevelt and accused him of lying. The *Tribune* did, however, print Roosevelt's letter to the editor, which had evoked the *Sun's* editorial, on a back page of their paper.

11. Bonaparte to Roosevelt, December 8, 1908.

12. Ibid. Italics in original.

13. Richardson, *Politics of Conservation*, 56.

14. Ibid., 56-57, 67, 69, 75, 83, 84, 112; Henry F. Pringle, *The Life and Times of*

William Howard Taft (New York, 1939), 386, 441, 494, 604, 656-77. Five of Taft's cabinet officers were former corporation lawyers.

15. George W. Wickersham to Ligon Johnson, October 6, 1910, DJ File 144276. On Ballinger and the others, see Richardson, *Politics of Conservation*, 61-63; "Big Business," chap. 34 in Pringle, *Life and Times of Taft*; and Stanley Donald Solvick, "William Howard Taft and the Progressive Movement: A Study in Conservative Thought and Politics" (Ph.D. diss., University of Michigan, 1963), 303-6.

16. Charles J. Bonaparte to George W. Wickersham, March 18, 1909, Bonaparte Papers; W. Murray Crane to T. H. Carter, July 9, 14, 1910, Carter Papers. Winthrop Murray Crane was a prominent Republican and an intimate of Theodore Roosevelt. He exerted important influence on Roosevelt yet declined a number of cabinet posts offered by the president. He was owner of a number of manufacturing and paper mills and supplied the government with the paper on which U.S. bank notes and treasury certificates were printed. He had been lieutenant governor and governor of Massachusetts (1897–1902) and as senator played a prominent role in foreign policy matters. He was deeply interested in the League of Nations. See Morison, *Letters of Theodore Roosevelt*, 6:520, 606, 845, 865-66; Henry F. Pringle, *Theodore Roosevelt: A Biography* (New York, 1931), 268-78, 340.

17. H. M. Hoyt to George W. Wickersham, March 13, 1909; George W. Wickersham to William Taft, June 1, 1909; M. M. Hays to George W. Wickersham, July 18, 1910; all in DJ File 144276.

18. Ligon Johnson to Attorney General, February 13, 1909, DJ File 144276.

19. Ibid.

20. Ibid.

21. Ibid.

22. C. F. Kelley to Ligon Johnson, January 7, 1909; Charles J. Bonaparte to Theodore Roosevelt, December 8, 1908; H. M. Hoyt to George W. Wickersham, March 13, 1909; all in DJ File 144276.

23. Johnson to Attorney General, February 13, 1909.

24. Ibid.

25. Ligon Johnson to George W. Wickersham, February 19, 1909, DJ File 144276.

26. Ibid.

27. Ibid.

28. Ibid.

29. Ibid.

30. Ibid.

31. Ibid.

32. Ibid. Johnson's letter to Kelley is quoted in its entirety in the report of February 19, 1909.

33. The form of the arrangement, excluding Kelley's concluding clause, was as follows: That the bill by the Government and this agreement shall be filed simultaneously and,

if agreeable to the Court, no public notice or publicity shall be given to the bill and this agreement or any other action be taken in connection therewith, beyond placing the bill in the official files of the Court, during the period covered by this agreement.

> That the scientific and commercial feasibility of the manufacture of sulphuric acid and fertilizer shall be investigated as early as practicable, the Company cooperating with the Government to this end.
>
> That as soon as weather conditions admit, the character and extent of the Western phosphate beds shall be ascertained.
>
> Should it be determined that the manufacture of sulphuric acid is scientifically and commercially feasible, the Company shall further be given a reasonable time in which to erect any necessary plant and appliances.

34. Ibid.

35. Ibid.

36. *Standard*, April 21, 1909; Ligon Johnson to Gifford Pinchot, November 10, 1909, Pinchot Papers; Attorney General to Nathaniel P. Pratt, June 12, 1909, DJ File 144276.

37. Pratt to Attorney General, June 12, 1909, DJ File 144276. White arsenic is commercial arsenic.

38. Bonaparte to Roosevelt, December 8, 1908.

39. The original documents in the suit are filed at the Federal Records Center, 6125 Sand Point Way, Seattle, Washington 98101.

40. *Standard*, March 17, 1910; March 7, 1911.

41. R. L. Clinton to Ligon Johnson, July 18, 1910; Ligon Johnson to George W. Wickersham, October 4, 1910; both in DJ File 144276.

42. Johnson to Clinton, July 18, 1910.

43. Ligon Johnson to W. J. Hughes, July 20, 1910, DJ File 144276.

44. *Standard*, December 5, 17, 1908.

45. T. H. Carter to Mrs. T. H. Carter (wife), June 12, 1910; W. Murray Crane to T. H. Carter, July 9, 1910; Charles D. Norton to T. H. Carter, July 12, 1910; and W. Murray Crane to T. H. Carter, July 14, 1910; all in Carter Papers.

46. *Standard* and *Miner*, December 5, 1908; *Miner*, March 15, 16, 1910; *Butte Evening News*, December 20, 1908; R. L. Clinton to William Taft, March 4, 1911, DJ File 144276.

47. Hedgcock, "Report on Forest Conditions . . . Anaconda, Montana." For a period of four or five years, the government tried to hold the report secret in the file against Anaconda to use in the anticipated litigation.

48. Ibid., 12-15. "A careful microscopic examination of specimens taken from each tree in each group examined from the Anaconda region failed to reveal the presence of parasitic fungi or bacteria in the leaves," Hedgcock noted. He concerned himself mainly with what he considered the five most important species of coniferous trees in the

Anaconda region, viz., alpine fir, Douglas fir, lodgepole pine, Engelmann spruce, and limber pine. Regarding the extent of injury to these species, he wrote: "In the region adjacent to Anaconda, less than five miles away, these species, with the possible exception of a few limber pines, have been nearly all denuded of their leaves and are dead in most instances." The various species of juniper were the most resistant. See also Fred Morrell to William B. Greeley, June 30, 1924, DJ File 144276.

49. Ligon Johnson to George W. Wickersham, October 8, 1910, DJ File 144276.

50. Peirce, Swain, and Mitchell to Johnson, October 21, 1910. Like the Hedgcock report, the Justice Department tried to keep this file secret in anticipation of filing suit against the Anaconda Company. As late as 1914 the government maintained that the report should remain confidential, and the professors were refused permission to publish the report and any of the other conclusions coming out of the continuing laboratory investigations. See Johnson to Hughes, June 20, 1913; George Peirce, Robert Swain, and J. Pearce Mitchell to Ernest Knaebel, July 1, 1914; both in DJ File 144276.

51. Peirce, Swain, and Mitchell to Johnson, October 21, 1910, 4. In such spots they found lodgepole pine and red firs.

52. Ibid., 7.

53. Ibid., 12.

54. Ibid., 14, 18. Further study and experiments on their findings were continued at Stanford.

55. C. F. Kelley to Ligon Johnson, October 4, 1910, DJ File 144276.

56. Ligon Johnson to George W. Wickersham, December 1, 1910, DJ File 144276.

Chapter 11: "We Will Have Secured All That We Have a Right to Ask"

1. Ligon Johnson to George W. Wickersham, December 13, 1910, DJ File 144276.

2. See John D. Ryan to Thomas H. Carter, December 20, 1910, Carter Papers; Johnson to Wickersham, December 13, 1910.

3. George W. Wickersham to Ligon Johnson, December 15, 1910; Ligon Johnson to George W. Wickersham, December 17, 1910; both in DJ File 144276.

4. The McCune cutting was eventually settled for $150,000 in February of 1914. Johnson himself concluded the settlement and in a letter to the solicitor general reflected on it in terms of how large the compensatory damages amounted to as a result of the Washoe's fumes. "The area embraced in the McCune cutting constitutes but a small part of that involved under the Government's claim for fume damage. I cite the McCune cutting to afford some idea of the amounts involved." Ligon Johnson to George W. Wickersham, January 2, 1911; Ligon Johnson to Solicitor General, February 17, 1914; both in DJ File 144276.

5. George W. Wickersham to Ligon Johnson, January 4, 1911, DJ File 144276.

6. George W. Wickersham to C. F. Kelley, March 20, 1911, DJ File 144276.

7. *United States v. Anaconda Copper Mining Company*, Stipulation, filed May 1, 1911; and *United States of America v. Anaconda Copper Mining Company*, Resolution of Trustees, Federal Records Center, Seattle, Washington.

8. "If such vacancy be caused by the death or resignation of John Hays Hammond, or his successor, appointed as herein provided, by joint agreement of the parties; and in case they shall fail to agree for a period of sixty days after such vacancy shall occur, then by the senior Circuit Judge of the United States for the Ninth Circuit. If caused by the death or resignation of J. A. Holmes, or his successor, appointed as herein provided, then by the Attorney General of the United States. If caused by the death or resignation of Louis D. Ricketts or his successor, appointed as herein provided, by written designation by the President for the time being of the said defendant company, lodged with the Department of Justice." *United States v. Anaconda Copper*, Stipulation.

9. The Anaconda Company agreed to pay to each of the members of the board for their travel and "a reasonable compensation for their services," not exceeding "one hundred dollars per day for time actually expended for work under this agreement." The company also agreed to compensate any experts hired by the chairman of the review board. Ibid.

10. Ibid.

11. Ligon Johnson to W. J. Hughes, August 21, 1914; Ligon Johnson to the Solicitor General, February 17, 1914; both in DJ File 144276. On the advice of Gifford Pinchot and J. K. Haywood in December 1908, Attorney General Charles J. Bonaparte had assumed damages to be a part of the case, in addition to injunctive relief. See Bonaparte, Memorandum as to Conference with President, December 8, 1908.

12. George W. Wickersham to Ligon Johnson, April 21, 1911; Ligon Johnson to George W. Wickersham, April 22, 1911; both in DJ File 144276.

13. *Standard*, May 2, 1911; *Who Was Who in America*, 5:513. For Hammond's role in the Jameson raid and his association with Cecil Rhodes, see Jean Van der Peel, *The Jameson Raid* (London, 1952), and E. H. Pakenham, *Jameson's Raid* (United Kingdom, 1960).

14. Isaac F. Marcosson, *Anaconda* (New York, 1957), 256-59; *Standard*, May 2, 1911; *Who Was Who in America*, 1:1032-33.

15. Further information on this will be found in chap. 12.

16. George W. Wickersham to Ligon Johnson, December 13, 1911, DJ File 144276.

17. J. Clarence Davies III, *The Politics of Pollution* (New York, 1970), 103; Fulton, "Metallurgical Smoke."

18. Fulton, "Metallurgical Smoke," 7.

19. George W. Wickersham to Secretary of the Interior W. L. Fisher, March 12, 1912, DJ File 144276.

20. Memorandum Regarding Trip to West by W. J. Hughes, August 19, 1908; Ligon Johnson to George W. Wickersham, February 11, 1914; J. B. Chatterton to Commissioner, General Land Office, August 23, 1912 (Confidential: Not for Public Inspection); Ligon Johnson to George W. Wickersham, November 11, 1911; all in DJ File 144276.

21. Johnson to Wickersham, February 11, 1914; Lambeth E. Gibson to Department of Agriculture, August 9, 1909; Chub Gibson to W. R. Harr, November 11, 1912; Ligon Johnson to George W. Wickersham, November 23, 1912; Lambeth E. Gibson to John E. Raker, June 10, 1913; all in DJ File 144276. Gibson writing on behalf of the Shasta County Farmers' Protective Association suggested to Raker that the federal government take over the smelting industry, establish smelters with "smoke consuming plants," do all the smelting, and charge each mining company a certain percentage per ton.

22. Lambeth Gibson to Wade H. Ellis, September 8, 1909; Lambeth Gibson to Department of Agriculture, August 9, 1909; both in DJ File 144276.

23. Ligon Johnson to George W. Wickersham, June 16, 1911; Memorandum for the Attorney General, June 20, 1911; George J. Peirce to Ligon Johnson, October 11, 1911; all in DJ File 144276.

24. Johnson to Wickersham, June 16, 1911; Peirce to Johnson, October 11, 1911.

25. See copy of suits, undated, against each of the smelters; and Ligon Johnson to George W. Wickersham, November 8, 1911, August 22, 1912; all in DJ File 144276.

26. Fulton, "Metallurgical Smoke," 13. (Among the cities cited were Berlin, with SO_2 measurements of 0.000035; London, 0.00039; and Glasgow, 0.00042.) Johnson to Wickersham, November 8, 1911; *Salt Lake Tribune*, November 15, 16, 17, 19, 1908; *Salt Lake Intermountain*, November 23, 1908.

27. Fulton, "Metallurgical Smoke," 82-83.

28. Johnson to Wickersham, November 8, 1911. The baghouse system, which involves filtering the smoke stream through woolen or cotton bags, was commonly used in most lead smelters because of health and legal requirements—and because fume from lead smelting is valuable. The fume from copper smelting was not as valuable as that from lead. The degree of success yielded by the baghouse process is dependent on the nature of the ore processed and the maintenance practices employed. Maintenance, Charles H. Fulton pointed out in his article of 1915, was expensive. See Fulton's "Metallurgical Smoke," 44-58, for the pros and cons of the baghouse method of fume recovery vis-à-vis other methods.

29. Johnson to Wickersham, November 8, 1911, in which he quotes Cottrell's article.

30. Fulton, "Metallurgical Smoke," 7, 59-60; Johnson to Wickersham, November 8, 1911. Fulton provided an analysis of the problems associated with the Cottrell method of recovering fume and dust. He felt that, given further experimentation, the method would largely solve the emissions problem.

31. Johnson to Wickersham, November 8, 1911.

32. Johnson to Wickersham, August 22, 1912.

33. Ligon Johnson to George W. Wickersham, August 22, 1912; November 7, 1911; August 21, 1914; all in DJ File 144276.

34. Ibid.; Ligon Johnson to George W. Wickersham, July 18, 1912, DJ File 144276. See also Fulton, "Metallurgical Smoke," 72-76.

35. Johnson to Wickersham, November 7, 1911.

36. Ibid.; George W. Wickersham to Ligon Johnson, November 14, 1911; George W. Wickersham to John Hays Hammond, November 16, 1911; Ligon Johnson to George W. Wickersham, July 18, 1912; all in DJ File 144276.

37. Johnson to Wickersham, July 18, 1912.

38. Ligon Johnson to George W. Wickersham, December 19, 1912, DJ File 144276; Fulton, "Metallurgical Smoke," 72-77, provides a comprehensive description of this process, cautiously analyzing both the "dry process" and the "wet process" to which the operation could be tailored.

39. Johnson to Wickersham, August 21, 1914; Fulton, "Metallurgical Smoke," 70-73.

40. Johnson to Wickersham, July 18, 1912.

41. Chatterton to Commissioner, General Land Office, August 23, 1912.

42. Ligon Johnson to George W. Wickersham, September 25, 1912; Ligon Johnson to Attorney General, August 1, 1913; Department of Justice Memorandum, June 29, 1914; all in DJ File 144276.

43. Johnson to Wickersham, September 25, 1912.

Chapter 12: The Struggle Abandoned

1. John Hays Hammond to Attorney General, December 13, 1912, DJ File 144276.

2. Ibid.

3. Arthur E. Welles, Report of the Anaconda Smelter Commission Covering the Period May 1, 1911, to October, 1920, Records of the U.S. Bureau of Mines, Box 84B, General Files to 1954, Record Group 70, National Archives, Washington, D.C.; Hammond to Attorney General, December 13, 1912.

4. Fulton, "Metallurgical Smoke," 7; Ligon Johnson to Solicitor General, February 17, 1914.

5. In 1915 the Amalgamated was dissolved under pressure from the federal government, and the name Anaconda once again became the trade name for the entire organization. See Toole, *Twentieth-Century Montana*, 122; and Burlingame and Toole, *History of Montana*, 1:279 n. 6.

6. A typical example is the following: "Construction work began July, 1917 [on building a bigger stack] when a wagon road and a considerable length of railroad was built to enable the company to deliver the materials right on the site of the installation." Welles, Report of Anaconda Smelter Commission, 1920, 65.

7. In late 1914 Johnson explained his new position to W. J. Hughes of the Justice Department, who was still the attorney in charge of the smelter cases: "My arrangement with this company is $1500 per month when away from New York and $1250 when in New York." Ligon Johnson to W. J. Hughes, August 21, 1914, DJ File 144276.

8. For Ligon Johnson's speech, see *AIME,* 58 (1925), 58, 198.

9. Welles, Report of Anaconda Smelter Commission, 1920, table 3, 38.

10. Ibid., 13.

11. Ibid., 75.

12. Ibid., 86-87, 76.

13. W. R. Roberts to Secretary of Agriculture, January 18, 1914, DJ File 144276.

14. Ibid.

15. J. T. Roseborough to Secretary of Agriculture, February 16, 1914, DJ File 144276. See also the correspondence of Harry J. Quinlan in DJ File 144276.

16. Department of Justice to J. T. Roseborough, February 26, 1914, DJ File 144276.

17. Chub Gibson to Harr, November 11, 1912; Gibson to Raker, June 10, 1913; George W. Wickersham to Gibson, November 4, 1912; Ligon Johnson to Attorney General, November 23, 1912; John E. Raker to Attorney General, n.d., 1912; W. J. Hughes to Ligon Johnson, June 23, 1913; Ligon Johnson to W. J. Hughes, July 13, 1913; all in DJ File 144276.

18. See the letters of John E. Raker to Attorney General, n.d., 1913; Hughes to Johnson, June 23, 1913; Johnson to Hughes, July 13, 1913.

19. Johnson to Hughes, July 13, 1913.

20. Ibid.

21. Fulton, "Metallurgical Smoke," 86. The same procedure was followed in the controversy over the Selby smelters alleged to be damaging lands in Benicia County near San Francisco. See J. A. Holmes et al., "Report of the Selby Smelter Commission," *Bureau of Mines Bulletin*, 98 (1915).

22. See the map of Anaconda's mining empire in "Anaconda," *Fortune Magazine*, January 1937. The writer discusses how Anaconda Copper had achieved its dominance in the domestic and world copper market by 1929, the year he said was the financial "apotheosis" of Anaconda, and had reached an industrial integration similar to what U.S. Steel had achieved decades earlier. "The chief difference between the nation's No. 1 copper corporation and the nation's No. 1 steel corporation [arises] from the fact that Anaconda sits astride the complicated world copper situation. Anaconda is a copper buyer domestically," the writer said, "a copper seller internationally. But in the end it is a fabricator."

23. Welles, Report of Anaconda Smelter Commission, 1920, 63-67.

24. C. F. Kelley to Van H. Manning, August 27, 1918, DJ File 144276.

25. Ibid.

26. Report of Anaconda Smelter Commission, May 1, 1924, DJ File 144276.

27. "Anaconda," *Fortune Magazine*, December 1936, 93-94. See also Louis Levine, *The Taxation of Mines in Montana* (New York, 1919), chap. 6.

28. "Anaconda," *Fortune Magazine*, December 1936, 94, 210.

29. Levine, *Taxation of Mines*, 38-48.

30. Welles, Report of Anaconda Smelter Commission, 1920, 35, 37.

31. Ibid., 97.

32. Ibid., 67-68.

33. Ibid., 68.

34. Ibid., 106, 119.

35. Report of Anaconda Smelter Commission, 1924.

36. Ibid., 7-8.

37. Ibid., 9, 11.

38. Ibid., 13.

39. Ibid., 16.

40. Ackley, Holden, and Magill, *Air Pollution Handbook*, 2-21.

41. In 1947–48 a study done by the U.S. Public Health Service found that the annual lung cancer death rate among males in Deer Lodge County was 145.7 per 100,000 as compared with 5.2 per 100,000 in Gallatin County, an agricultural area to the southeast. Dr. W. C. Hueper of the National Cancer Institute, an established authority on environmental cancer, said of the above data, that "the high incidence of lung cancers among the population of several counties in Montana where copper smelters and mines were operated for many years creating an occupational and environmental pollution of the atmosphere and soil with arsenicals" strongly suggests that "Prolonged inhalation of arsenical dust and fumes appears to produce an increased liability to cancer of the lung." See W. C. Hueper, M.D., "A Quest into the Environmental Causes of Cancer of the Lung," *Public Health Monograph No. 36* (1956), published by the U.S. Department of Health, Education and Welfare. See also Montana State Board of Health, "A Study of Air Pollution in Montana, July 1961–July 1962," n.d., on file at the University of Montana Library, Missoula.

42. See Hubert Work to Irvine L. Lenroot, January 31, 1924, *Senate Reports on Public Bills*, 68th Cong., 2d sess., 1925, S. Rept. 1267, serial 8388 (hereafter *Senate Reports*); Haywood, "Injury . . . by Smelter Wastes."

43. W. C. Staton to Ligon Johnson, July 18, 1910, DJ File 144276.

44. In proceedings of this nature, only in the case in which the court deems a "substantial injustice" has been done in either procedure or interpretation of the law can a revision in the decision of the case under review be granted. See *Black's Law Dictionary*.

45. R. L. Clinton to George W. Wickersham, February 21, 1911; memorandum by President Taft to George W. Wickersham on behalf of R. L. Clinton, February 21, 1912; George W. Wickersham to R. L. Clinton, February 23, 1912; memorandum by President Taft's Secretary to George W. Wickersham, February 26, 1912; Harry J. Quinlan to Woodrow Wilson, July 21, 1913; Harry J. Quinlan to Attorney General, October 28, 1913; Solicitor General to Harry J. Quinlan, November 3, 1913; Harry J. Quinlan to Woodrow Wilson, November 13, 1913; James D. Maher to Harry J. Quinlan, telegram, January 21, 1914; Harry J. Quinlan to Solicitor General, January 21, 1914; Solicitor General to Harry J. Quinlan, February 2, 1914; Solicitor General to Harry J. Quinlan, November 22, 1915; Fred Morrell to William B. Greeley, January 14, 1924; Ligon Johnson to Greeley, March 4, 1924; all in DJ File 144276.

46. Report of Anaconda Smelter Commission, 1924, 9.

47. Between those years the Interstate Commerce Commission endeavored to regulate the railroads. The Supreme Court heard sixteen cases under the Interstate Commerce

Act. Of that total, it found for the railroads in all but one case. William Z. Ripley, *Railroads: Rates and Regulations*, vol. 1 (New York, 1912), 463.

48. Marcosson, *Anaconda*, 153-58.

49. John Kenneth Galbraith, *The Great Crash, 1929* (Boston, 1961), 6-9.

50. J. R. Hobbins to Fred Morrell, October 9, 1923, Anaconda "Smelter Fumes" Exchange File, U.S. Forest Service Regional Office, 200 E. Broadway, Missoula, Montana.

51. Ibid.

52. Memorandum for Mr. Riter, November 1, 1923; William B. Greeley to Fred Morrell, October 20, 1923; both in DJ File 144276. See *Congressional Record*, 68th Cong., 2d sess., 1925, 66, pt. 5:5259, and *Senate Reports*.

53. See Henry C. Wallace to Irvine L. Lenroot, January 21, 1924; Hubert Work to Irvine L. Lenroot, January 31, 1924; both in *Senate Reports*.

54. Wallace to Lenroot, January 21, 1924; Work to Lenroot, January 31, 1924. Senator Walsh's papers at the Library of Congress do not disclose his motives. See Thomas J. Walsh Papers, Library of Congress, Washington, D.C.

55. Forest Service, "National Forest Areas," June 30, 1925, at the University of Montana Forestry Library. For the Clark-McNary bill and the Forest Service's analysis of it, see William B. Greeley, *Report of the Forester*, June 30, 1924, also at the University of Montana Forestry Library.

56. Fred Morrell to William B. Greeley, January 14, 1924, DJ File 144276.

57. William B. Greeley to Ligon Johnson, February 11, 1924; Ligon Johnson to Greeley, March 4, 1924; L. F. Kneipp to Solicitor, Department of Agriculture, March 5, 1924; W. Marvin to Harlan F. Stone, March 20, 1924; all in DJ File 144276.

58. Harlan F. Stone to Henry C. Wallace, May 28, 1924, DJ File 144276.

59. *Congressional Record*, 68th Cong., 2d sess., 1925, 66, pt. 5:5259; *Senate Reports*.

60. Roughly computed from U.S. Department of Commerce, *Historical Statistics of the United States*, 1960, 310. A table provides statistics on national forest areas and purchases between 1905 and 1957, with the average price per acre paid for the total area purchased each year.

61. The exchange of 1937 amounted to only $29,380. See the Anaconda "Smelter Fumes" Exchange File, U.S. Forest Service Regional Office, Missoula, Montana, or Federal Records Center, Seattle, Washington. See also *United States of America, Complainant, v. Anaconda Copper Mining Company and the Washoe Copper Company, Defendants*, filed and entered March 16, 1910, Federal Records Center, Seattle, Washington.

62. Office of the Clerk of the Federal District Court, Butte, Montana. The records in the case do not show that a final release was ever executed so as to terminate the litigation in the manner demanded by the vice president of Anaconda Copper in his letter to the district forester in Missoula. The action by the court clerk was an administrative procedure. It was not an adjudication of the case.

– Index –